D1077357

5 Au.

Jr 2i

Dec '2

# DO WE NEED PANDAS?

C014693132

# DO WE NEED PANDAS?

## THE UNCOMFORTABLE TRUTH
## ABOUT BIODIVERSITY

**Ken Thompson**

green books

First published in 2010 by

**GREEN BOOKS**
Foxhole, Dartington
Totnes, Devon TQ9 6EB
www.greenbooks.co.uk

© **Ken Thompson 2010**
All rights reserved

Illustrations © Jennifer Johnson

Design by Stephen Prior


Printed in the UK by TJ International, Padstow, Cornwall.

The text paper is made from 100% recycled post-consumer waste;
the covers from 75% recycled material.

**DISCLAIMER:** The advice in this book is believed to be correct at
the time of printing, but the author and the publishers accept no
liability for actions inspired by this book.

ISBN 978 1 900322 86 7

| Hampshire County Library | |
|---|---|
| C014693132 | |
| **Askews** | Sep-2010 |
| 333.9516 | £9.95 |
| | 9781900322867 |

# Contents

# Acknowledgements

This book borrows from the ideas of hundreds of scientists, to all of whom I'm deeply grateful, but it would be impossible to mention them all. Indeed, I'm determined not even to try, because I know that once I start I will only worry about who I've managed to leave out. Nevertheless, I must acknowledge that this book could not have been written without Phil Grime FRS, my long-suffering mentor ever since my first serious steps as a scientist. Phil's actual science crops up only intermittently in the book, and I wouldn't presume to think that he would agree with all of it, but his influence is on every page.

I considered, but rejected, providing a detailed reference list. Almost all the ideas and information in the book come from the primary scientific literature, i.e. papers published in scientific journals; around 200 of them in all. This is not, frankly, a literature that was written with the understanding or enjoyment of the general public as its principal objective. Nevertheless, for readers who wish to follow up some of the sources, a small number of key references are identified in the text and listed at the back of the book. For those who want some more general reading about biodiversity, I recommend *Biodiversity: A beginner's guide* by John Spicer (2006, Oneworld Publications) and *The Diversity of Life* by E. O. Wilson (2001, Penguin). For an unparalleled but accessible survey of the nuts and bolts of biodiversity, I also recommend *The Variety of Life: A survey and a celebration of all the creatures that have ever lived* by Colin Tudge (2002, Oxford University Press).

I owe a deep debt of gratitude to Mick Uttley, who undertook to read a draft of the entire book. The book is much improved as a result of his perceptive comments, although of course any errors and omissions remain entirely my own.

I wish to thank John Elford of Green Books, for having confidence in this book when I was just beginning to think no one ever would, and Alethea Doran for her expert copy-editing of the manuscript. I would also like to thank everyone else involved at Green Books, and Jennifer Johnson for her illustrations.

Finally, thanks to Pat for her love and encouragement during the writing of this book, and for prodding me when I sometimes didn't feel like getting on with it, and to Rowan for his usual unswerving support for my attempts at literature. Lewis: have I finally written a book you would consider reading?

# Foreword

Like any field of human endeavour, nature conservation, and the science that underpins it, can sometimes become weighed down with the baggage of its own complexity and convolutions of debate.

Jargon is a terrible curse. The word 'biodiversity' itself is arguably just a longer and more complicated way of saying 'nature', but it has gained in popularity because it sounds more objective and scientific somehow; more quantifiable and less prone to the vagaries of emotion.

This lack of comfort with subjectivity betrays an issue that cuts to the heart of the conservation challenge and the role of science. Sometimes we fight hardest to protect things because they move us, not because of objective valuation of their functional importance. That's why conservation charities use images of species like pandas for fundraising rather than, say, bacteria. Its a completely legitimate way to choose priorities, but it's also true that the world's health and our own survival depend on saving things that we sometimes find ugly and hard to empathise with. Bacteria keep us alive more than bears do, but the bears help us to have lives worth living.

Getting the balance between these different perspectives is far from easy, and it's even hard for scientists, who sometimes let their love for the life, the places, the ideas they study turn into a fierce belief in theories that are far from proven.

We betray our world views in the language we use. Visions of nature as a fragile, balanced, harmonious web paint a picture of a world of exquisite beauty but also one that is easily wrecked and destabilised. But nature is also adaptive, resilient, vigorous and regenerative. Often it has a surprising ability to recover from disturbance, but not always in the patterns and with the species we believe 'should be there naturally'. Sometimes nature seems hell-bent on doing things that don't strike us as natural.

Humans have launched a series of environmental changes that will reverberate across the world for centuries to come. A balanced, harmonious world no longer seems to be an option open to us. To understand the impacts on the nature we depend on, we need to get better at telling the difference between ideas that are true, those that we wish were true and those that we wish to make come true.

Which is why Ken Thompson has produced something that is not just a great book, it's an important book. The gentle, warm, straightforward prose walks us easily through many of the fundamental principles and questions that lie behind conservation science. But his writing is underpinned always by a clarity of thought that is rare and that helps us all understand why some things matter, why some things don't and why sometimes we just don't know.

There aren't many things more important than understanding how to sustain the world that sustains us. Ken Thompson has written a perfect book for everyone who wants to understand more about the world we live in.

Tony Kendle
Foundation Director, Eden Project

# Introduction

Turn on the television, pick up a newspaper or surf the internet, and you can hardly avoid news about biodiversity, nearly all of it bad. Yet most of this coverage still manages to avoid the really important questions that most of us would like answered: what is biodiversity; where is it to be found; why is it threatened; and what, if anything, can we - or should we - do to save it? Perhaps most crucially, what does biodiversity actually do, and how alarmed should we be about its continuing loss? I don't guarantee to provide complete answers to any of these questions, but I hope that by the end of this book you will have a better idea of what we know about biodiversity, and - perhaps more importantly - a better idea of what we don't know.

This book has a strong botanical bias, especially in its early chapters. This might seem curious, given that most people will have been alerted to the perils facing biodiversity by reports of threats to charismatic animals such as pandas, whales and tigers. In mitigation, I make two pleas.

First, if we are to understand biodiversity, and how to save it, we must understand ecosystems - that is, the entire living world plus its non-living environment - and ecosystems consist overwhelmingly of plants. For example, plant biomass in a temperate woodland averages around 300,000kg per hectare, but the animals in the same area weigh only about 300kg (0.1 per cent of the plants) - and almost all of that is earthworms. Not only do plants make up all but a negligible fraction of the biomass, but they are where carbon, nutrients and energy enter ecosystems, and they provide both the physical habitat and the living and dead material that supports the animal food chain. The quantity, quality and diversity of vegetation is ultimately responsible for the character of the entire ecosystem. Given their pivotal role in maintaining life on Earth, plants don't always get the attention they deserve, for the simple reason that there are so many more animals, in terms of both individuals and species.

Second, I'm a botanist, for better or worse, and not only do I prefer to stick to things I know something about, I also happen to think that the science of biodiversity is more advanced for plants than for animals. Nevertheless, I urge those of you who like your wildlife to have fur or feathers to be patient; there are plenty of animals later in the book.

Dodo, *Raphus cucullatus*

## Chapter 1

# What is biodiversity?

*O wonder! How many goodly creatures are there here!*
**William Shakespeare, *The Tempest***

## How many species are there?

Since this book is about biodiversity, I suppose a good place to
start would be: what exactly is biodiversity? The Rio Convention on
Biological Diversity says:

> *Biological diversity - or biodiversity - is the term given to the variety
> of life on Earth and the natural patterns it forms. The biodiversity
> we see today is the fruit of billions of years of evolution, shaped by
> natural processes and, increasingly, by the influence of humans. It
> forms the web of life of which we are an integral part and upon which
> we so fully depend. This diversity is often understood in terms of the
> wide variety of plants, animals and microorganisms.*

In other words, on one level (the simplest), biodiversity is the number
of different kinds of living things on planet Earth.

So, how many living things are there? Well, one place to start estimating
is those species we have described, classified and named. This
amounts to about one-and-three-quarter million, over half of them
insects and less than a quarter of them plants in the strict sense of
the word. How close is this figure to the true number? It would be
unduly pessimistic of me to say we haven't a clue, but the truth is
not much better than that. However, there is no shortage of guesses.
One of the most celebrated was made by Terry Erwin in 1982, in what
must surely be the most famous paper ever published (or ever likely

to be) in *The Coleopterists Bulletin*.[1] Erwin was (and is) an expert on the beetles of tropical tree canopies, and in particular the genus *Agra*, in which he has named many new species; *Agra vation* and *Agra cadabra* to name but two. Erwin's research led him to suspect that there is a lot more biodiversity in the canopies of tropical trees than anyone realised. Using an insecticide 'fog', he extracted all the creatures from the canopies of 19 individuals of one particular kind of tropical tree in Panamá. Most of these animals were insects and most were new to science. Around 1,200 of the species that fell out of his sampled trees were beetles, and Erwin estimated that 163 of them live on only this one tree species. We know there are around 50,000 tropical tree species, so if we assume that our target tree is typical (in other words, if all tropical tree species have about the same number of unique beetles), then there must be just over eight million different beetles in the canopies of tropical trees. About 40 per cent of all arthropods (animals with jointed exterior skeletons: insects, spiders, etc.) are beetles, so that means there must be about 20 million tropical-canopy arthropods. Around half as many species again live in or on the ground, making 30 million tropical arthropods in all, most of them insects.

It's not difficult to find fault with Erwin's method, and indeed biologists have been arguing about it ever since. Supposing Erwin's method missed some of the beetles that live on the trees he sampled? In which case his estimate would be too low. What if he overestimated the proportion of specialists among the beetles he caught? In which case it would be too high. In short, it's easy to see why hardly anyone agrees with Erwin's estimate, and it may indeed be too high (or too low), but to argue about the exact number is to miss the point: there are *very* many more species than we have named, and we haven't much idea how many.

Of course, you might argue that tropical rainforest, hugely diverse and largely unexplored, is where you would expect to find most of the Earth's unknown biodiversity. And maybe you'd be right, but there are unknown and unrecognised species everywhere. For 15 years college lecturer Jennifer Owen trapped, identified and counted the animals in her suburban garden in Leicester, an average-sized city in the middle of England. Simply by looking at one family of parasitic wasps, a group that very few people care about and fewer still are qualified to identify, she found 15 species not previously recorded in Britain and four species completely new to science.[2] So great was the effort required that she and her wasp expert kept up the sampling and identification of this one family of insects for only three years. There's every reason to expect that more effort would have yielded more discoveries.

But insects, by and large, are rather small. Surely there's nothing *large* out there waiting to be discovered? Wrong again. Even today, a new large terrestrial mammal is discovered roughly every three years, while the ocean yields a new large fish or mammal every five years. Altogether, 408 new mammal species have been described since 1993, including a pygmy sloth from Panama, a 'giant' muntjac deer from Vietnam, a white titi monkey from Brazil, and the Solomon Islands' monkey-faced bat. Nor is it just new animals that continue to be found. An expedition in 2002 to northern Vietnam discovered a species belonging to a completely new genus of conifers, the golden Vietnamese cypress. In fact the locals had known about this tree all along, and were busy cutting it down for its valuable timber. The same expedition discovered more than a hundred other new plant species, including several shrubs and about two dozen orchids.

In 2008 botanists were startled by the discovery of an enormous new palm on the island of Madagascar. The palm, originally spotted by Xavier Metz, a Frenchman who manages a cashew plantation nearby, has a huge trunk 18m in height, topped by fan-shaped leaves 5m in diameter. It's the most massive palm ever found on the island and one of the largest known flowering plants. Ironically, once you know they are there, the trees can plainly be seen on Google Earth. Across the whole of life, around 300 new species are described every day, a rate that shows no sign of slowing.

Of course, you might say, we're back to the tropics again, which is surely where all the large undiscovered plants are? Wrong again. In 1994, David Noble, from the New South Wales National Parks and Wildlife Service, Australia, was walking in the Wollemi National Park, about 150km north of Sydney. He saw some trees he didn't recognise, took a sample home, and found he had stumbled on a completely new conifer - and no ordinary conifer either. The nearest relatives of the Wollemi pine (*Wollemia nobilis*) are the Norfolk Island pine, the monkey puzzle and other members of the Araucariaceae family. In truth, however, it doesn't have any really close living relatives. Its real relatives are fossils dating back 150 million years to the age of the dinosaurs. The Wollemi pine grows 35m tall, with elegant foliage and bark like no other tree, best described as looking like gently simmering chocolate. Despite being one of the world's rarest plants, numbering as few as 43 adult trees at the time of its discovery, it looks like following in the footsteps of that other great survivor, the ginkgo (*Ginkgo biloba*), and enjoying a completely new lease of life in cultivation.

## The diversity of the very small

But all this diversity, impressive as it is, is merely the tip of the iceberg. As organisms become smaller they become both more numerous and more diverse, and few things are as small or as numerous as · bacteria. There is a tendency to think that because of their importance to medicine and molecular biology, bacteria are reasonably well known. Nothing could be further from the truth. One hundred grams of ordinary garden soil contains 100 billion bacteria - as many as there are stars in the galaxy. Yet how many species of bacteria have been described and named? About 4,500. The reason is simple: the first step in studying bacteria is to isolate single species by growing colonies from individual cells on a culture medium, but far less than one per cent of bacteria can easily be cultured in this way. The rest just refuse to cooperate and therefore cannot be studied by conventional techniques. Biologists have begun to get round this problem by looking directly at bacterial DNA. The results are startling: one gram of soil contains 4,000-5,000 different kinds of bacterial DNA, and thus the same number of different bacteria, while one gram of marine sediment contains another, quite different 4,000-5,000 bacteria. (Don't think, by the way, that DNA is only of use for sorting out the taxonomy and diversity of the very small. Searching for elusive large creatures is much simpler now that an animal can be identified by a DNA profile from a single hair. If the yeti exists, you can bet we will become aware of its DNA before anyone catches the animal itself.)

There are other, even more cunning ways around the refusal of bacteria to stand up and be counted. Ecologists have discovered that communities of living organisms obey certain 'laws', or highly repeatable patterns of abundance of the species within them. Essentially, a few species are common and almost all are rare. If you know how many individuals are present, and how common the top few species are, a combination of some reasonable assumptions and some fancy mathematics allows you to make a stab at how many species there might be. It's a bit like guessing how many postmen, teachers and bus drivers live in a country or city, knowing only the total population and the proportion of bank managers and doctors; it should work as long as different professions occur in reasonably fixed proportions. If you do this for bacteria, one gram of soil probably contains between 6,400 and 38,000 species; in other words, the DNA estimate is in the right ballpark. Scientists are always reassured when two completely different methods of estimating an unknown quantity turn up with similar answers: there's always the possibility that one method might be completely up the creek, but two different methods are unlikely to be wrong in exactly the same way. Astonishingly, however, if you do something that can't be done in the lab and ask how many different kinds of bacteria are in a *tonne* of soil, the answer is around four million. How is that possible? The key to such huge diversity is the enormous physical diversity of the soil environment at the scale experienced by bacteria. A fragment of decaying leaf harbours one bacterial community, the clay particles next door have a quite different set, while yet another community inhabits the gut of a nearby earthworm. The total surface area of all the tiny pores and channels in two tablespoons of soil is about the same as a typical city block: for bacteria and other creatures that live on the surface of soil crumbs, there's an awful lot of room in soil. This is in sharp contrast with the world's oceans, which are essentially structureless and may contain 'only' two million species of bacteria altogether.

Even though bacterial diversity beggars belief, this is only part of the story. C. Northcote Parkinson once observed that since the British Civil Service (in defence of its policy of employing only classics graduates) was unable to decide whether one man was better in geology than another man in physics, it was convenient to rule them both out as useless. We have long had a similar attitude to bacteria. Since they are very small, prokaryotes (lacking a true nucleus), and for the most part cannot be grown in the laboratory, it has been convenient to lump them together as a single kingdom, in the same way that all plants (or animals) constitute a single kingdom. In reality, there may be ten or fifteen bacterial kingdoms, each at least as different from the others as animals are from plants. A new bacterial phylum (a phylum is the level of classification just below a kingdom, so arthropods - insects, spiders, crustaceans etc. - are a phylum) is discovered every month, and the further we look, the more we discover that there seems no limit to where bacteria can live. In 1993 living bacteria were found in rocks over two kilometres below the Earth's surface, where they may have survived for millions of years without any contact with the surface. Nor is all this hidden diversity confined to the bacterial realm. The microscopic single-celled animals that were once erroneously lumped together as protozoa may be equally diverse. And the diversity of some other living organisms is turning out to be much higher than anyone thought, simply because many of these species live in places nobody thought to look in until recently. We've known for a long time that plants form symbiotic partnerships with various fungi that live in or on their roots. This *mycorrhizal* symbiosis seems to be crucial for most plants to obtain enough phosphorus from soils. It's also very ancient: fossils of the earliest land plants show clear signs of mycorrhizal symbiosis.

But what no one realised until recently is that there are many other fungi associated with plants. Every plant species so far examined (although not every individual) plays host to fungi that live *inside* its leaves. These fungal *endophytes* may be remarkably diverse; one estimate being that there are as many different fungal endophytes as all other fungi combined. Like many other groups, the pinnacle of their diversity seems to be in tropical forests. In Puerto Rico, up to 17 species of fungal endophyte were found inside a single leaf, each occupying an area of only a few square millimetres. The obvious question is: What are all these fungi doing? For nearly all of them the answer is: we've no idea. Many may have no obvious effects on their hosts at all, but we know that some contain toxins that help to defend the plant against predators and pathogens, while others seem to increase tolerance of salinity or drought. One American grass can grow at soil temperatures up to 65°C next to hot springs in Yellowstone National Park, but only if infected with a particular endophyte. What's more, it turns out that the endophyte can only confer this remarkable heat tolerance if it is itself infected with a certain virus. Turf and grass merchants have not been slow to appreciate the commercial possibilities, and are already marketing 'endophyte-enhanced' grass cultivars for extreme environments.

# A lot more than fish in the sea

By now many readers will have noticed that so far, with the arrogance typical of a terrestrial primate, I have concentrated almost exclusively on organisms that live on dry land. The simple, pragmatic reason is that we still know relatively little about life in the sea. More people have walked on the moon than have visited the deepest oceans, and we have better maps of the surface of Mars than of the ocean floor. In 2005 a US submarine crashed into a 2,000-metre submarine mountain that rises to within 50 metres of the surface near Guam. No one knew it was there. It's easy to forget that almost three-quarters of the area available for life, and about 99 per cent of the actual space, is in the sea. If all dry land down to sea level were bulldozed into the ocean, it would occupy just one-twenty-third of the ocean volume. To a mere landlubber most of the sea looks the same, but the oceans harbour habitats as distinct as are tropical rainforests, dry deserts and the polar ice caps. The diversity and sheer variety of life in the sea is correspondingly enormous. Many animal phyla (the plural of phylum) are entirely marine, and it's no accident that the first port of call for anyone wanting to appreciate animal diversity is the nearest rocky shore. Dry land, totally dominated by insects, is exceedingly dull in comparison. And yet, especially in the case of the deep ocean, marine biodiversity remains almost unknown.

Among the many startling animals down there is the siphonophore *Praya dubia*, the largest invertebrate in the world (and indeed the longest animal of any kind, at up to 40m). Trailing a curtain of tentacles bearing powerful stings, this is truly one of the planet's great killing machines. In fact the deep sea quickly teaches you that you can forget almost everything you thought you knew about what it takes to be a predator. In the deep ocean 'jellyfish' (in the sense of organisms made of jelly) rule: animals made of 95 per cent seawater and just enough skin, muscle and nerves to stop them falling apart. So fragile that most are impossible to catch in one piece; often quite transparent, without teeth, bones or brains; they are nevertheless the dominant predators of the open ocean. Of course there are plenty of vertebrates too, some of them extraordinarily abundant. The deep-sea benttooth bristlemouth fish (*Cyclothone acclinidens*) may be the commonest vertebrate in the world, yet only a handful of marine biologists have even heard of it, and fewer still have seen one (they are only 3-4cm long, and largely transparent). For the last 25 years a new deep-sea species has been discovered about once every two weeks, and the number yet to be discovered is estimated at anywhere between 10 and 30 million. So if you think I'm not paying enough attention to life in the oceans, then I apologise, but that's just a reflection of how little we know.

## Why are there so many species?

Counting the Earth's species is only the start. What are they all doing, and how do they manage to coexist? Of course, the basic answer to this question is natural selection. It was Charles Darwin's genius to recognise that the unique profligacy of the natural world makes it also uniquely creative. All over the world, in every generation, trillions of animals and plants produce many more offspring than can possibly survive. Mostly these offspring are not identical, although the differences between them may be only subtle. Some of these differences are merely skin deep, and are not heritable, but many are the result of tiny genetic mutations and as such can be inherited. At the same time, all genes are constantly being mixed and shuffled into new combinations by sex, massively accelerating the changes that could occur by mutation alone. Living organisms are therefore constantly probing the possibilities of making a living from the environment, by the simple expedient of keeping those that do this best and killing off those (the overwhelming majority) that aren't quite as good. Since this environment itself consists largely of other species, all doing the same, it's not hard to see how natural selection continuously creates both new opportunities and the organisms to exploit them. For the most part, evolution by natural selection proceeds at what appears, to human senses, to be a snail's pace, but every now and then it takes sudden and rather dramatic strides. Sometimes this is because some barrier is removed, as when the dinosaurs disappeared and left the field clear for the rapid evolution of the mammals. Sometimes evolution itself gives rise to an innovation that allows a new group to power ahead, as when the previously dominant cycads and conifers gave way before the proliferating flowering plants.

But natural selection alone is not enough. If all the world's biodiversity were like one homogeneous, frequently stirred pot, the power of natural selection to create new species would be seriously compromised. Populations at the edges of species' ranges are often exposed to rather atypical conditions, leading to a tendency to evolve characteristics that differ from those in the centre of the range. As long as these different populations are in contact with the main bulk of the species, however, it's very hard for them to evolve into new species, because their new features keep being swamped by the flow of genes from the centre. In reality speciation (i.e. splitting into two or more daughter species) works best, or perhaps only works at all, when populations of animals and plants are free to evolve in isolation. In fact, so effective is isolation at promoting speciation that it can occur even when the conditions experienced by the isolated population are identical to those experienced by the parent species. Isolation of just a small part of the original gene pool (the 'founder effect'), or the loss of key genes by chance ('genetic drift') can lead quickly to the evolution of a new species.

The results can be seen on many levels. On a geological timescale, the northern hemisphere was for a long time one single land mass, which broke up into two chunks (Eurasia and North America) relatively recently, so it has essentially a single flora. If you are familiar with the plants of Europe, you won't find too many surprises among the floras of North America or Japan. Each region has different but related oaks, beeches and maples. Sometimes a piece of the jigsaw is almost or quite missing. For example, both North America and Asia have magnolias, and both are now grown in gardens in Europe, but Europe's magnolias never made it through the ice ages. The same thing almost happened to rhododendrons, with just six species now native to Europe out of more than 850 worldwide (although, just as with magnolias, British gardeners now grow plenty of Asian and a few American rhododendrons). On a smaller scale, Britain's flora was mostly wiped out by glaciations on several occasions in the last few million years, and recolonised after the ice retreated. Therefore it's essentially just a subset of the European flora, since the ice last retreated only 10,000 years ago. This is (just) long enough for Britain to have begun to evolve a few endemic species, i.e. species (in this case plants) that are found nowhere else. Patriotic botanists think Britain has quite a few endemics, while those with more inclusively European sympathies find only a few, but either way there's no denying that all of them have close relatives across the channel.

The Atlantic Ocean, the deserts of central Asia and the English Channel are obvious barriers, but in fact such barriers can operate at any scale. The evolution of finches and giant tortoises into different species on each of the Galapagos Islands, separated from each other by only a few kilometres of water, was one of the first clues that alerted Darwin to the existence of natural selection (ironically, he saw them as such compelling evidence for his new theory of natural selection only much later, and neglected to label each specimen with the island it came from). In fact island archipelagos are particularly fertile ground for natural selection, on account of the tendency of island species to abandon the ability to disperse. If you live on a small island, the power to disperse more than a few kilometres is of little value, since movement of this distance in any direction simply lands you in the sea. All island organisms rapidly tend to become sedentary stay-at-homes, thus reducing contact with populations on other islands and leading directly to the evolution of new species.

Island birds in particular show a strong tendency to evolve flightlessness: the classic example being arguably the world's most famous extinct animal, the dodo. Despite its curious appearance, the dodo was merely a pigeon that got stuck on an island and evolved large size and flightlessness. Research in 1996 showed that plants on islands can begin to evolve reduced dispersal ability within just a few generations. One of the most spectacular and well-documented examples of island speciation is the endemic Hawaiian fruit flies of the genus *Drosophila*. Around 1,000 species have arisen from what is believed to have been a single introduction about 26 million years ago. The flies have repeatedly hopped from older islands to newly formed ones, becoming cut off from their parents and evolving into new species along the way. Not only that, but fly populations on single younger islands are repeatedly fragmented by lava flows, and this constant splitting into small, isolated populations has promoted the evolution of hundreds of species. Nor is rapid evolution confined to real islands, surrounded by sea. Many habitats are effectively 'islands'. Adjacent ridges in the Peruvian Andes may have quite different floras, which are effectively isolated from each other by the intervening valleys.

Isolation need not involve real spatial, geographical separation. Even plants and animals that live side by side can easily become separated genetically, and once two populations can no longer exchange genes, they have taken the first step on the road to becoming different species. In animals, tiny variations in behaviour or pheromone chemistry can prevent two populations from interbreeding. Populations of the same plant in two adjacent fields can become isolated from each other, if differing human management favours different flowering seasons. An impressive example is tropical orchids, which have tended to evolve highly specific partnerships with particular insect pollinators. These partnerships isolate orchid species from each other, because the pollinators are confined to single orchid species. But the capacity to interbreed remains, and has been spectacularly exploited by human breeders of orchid hybrids. Such 'reproductive isolation' can produce new species at a stroke, sometimes by the mutation of a single gene.

## Various challenges, many solutions

So is that it? Is natural selection, combined with various degrees of isolation, enough to account for the Earth's remarkable biodiversity? Almost, but one more crucial ingredient is required: trade-offs. Trade-offs are why an Olympic long-jumper will never win any medals in the discus; in other words, nobody (and no species) can be good at everything. The concept is easiest to grasp by looking at plants. For plants, the world is a surprisingly simple place, for every plant needs exactly the same handful of essential resources: light, water, carbon dioxide and a few mineral nutrients. Since every species needs these same few things, why haven't a few become very good at getting hold of them, leaving little or nothing for all the others? Why isn't the world covered with just a few all-conquering plants, terrestrial versions of the red weed from H. G. Wells' *The War of the Worlds*? Trade-offs are one of the principal answers to this conundrum. Animals also provide countless illustrations of the same phenomenon, but animals use such a wide variety of different resources that things are never quite as clear as they are in plants.

Consider the simple example of seed size. Anyone who has ever grown a plant from seed knows that not all seeds are the same size. Some plants have big seeds and others have very small ones. Big seeds have big advantages: as every gardener knows, the seedlings of large-seeded plants like runner beans need less pampering than the tiny seedlings of busy lizzies or pelargoniums. In fact, although big seeds are not without disadvantages, most plants would like to have big seeds if they could. The trouble is that every plant would also like to have lots of seeds, and it's not really possible to have both. Just as you may have to decide whether to buy a new car or a foreign holiday, plants have to allocate limited resources to competing priorities. A typical heather bush might manage to produce over a million seeds per year, but only because each seed weighs only 0.03 milligrams. An oak tree, despite being 50 times bigger than a heather bush, might produce only 50,000 acorns per year, because each is 100,000,000 times heavier than a heather seed. How plants position themselves on this seed size-number continuum is not really relevant here, but the consequences can be far-reaching. For example, one reason you rarely find heather growing in shade is that its seeds are far too small. To begin life where there isn't much light, it helps to have a big seed like an acorn. On the other hand, heather seeds have to hang around in the soil for years or decades, waiting to germinate after the fires that periodically sweep through heathlands. Big seeds can't do this, chiefly because they are just too attractive to too many predators. An acorn that tried to behave like a heather seed would simply end up as a meal for a squirrel or a mouse. Essentially oak and heather need their seeds to do rather different jobs, and one size does *not* fit all.

Another trade-off familiar to gardeners concerns soil pH. We all know that some plants prefer acid soils, and indeed real acid specialists, such as rhododendrons, won't grow anywhere else. Because acid soils contain dangerously high levels of metals, such as iron, manganese and aluminium, acid specialists have erected barriers to keep most of these metal ions out of the plant. But they can't switch this ability on and off, so when faced with more alkaline soils, where these metals are highly insoluble, plants of acid soils suffer from iron deficiency, familiar to gardeners as lime chlorosis. Conversely, plants of alkaline soils have to try hard to absorb any iron at all, leaving them at the mercy of aluminium and manganese toxicity on acid soils.

## The fundamental rules of plant economics

Although seed size and soil pH illustrate the trade-off principle very well, there is another, much more fundamental trade-off that ultimately governs the ecology of all plants. As every gardener knows, some plants are very good indeed at acquiring the materials needed to grow quickly, and where resources are plentiful, such plants can exclude most potential competitors. Many natural monocultures consist of such plants, e.g. willows, poplars, brambles, Japanese knotweed, nettles and common reed. However, these capitalists of the plant world, expert at rapidly investing their captured materials in more leaves and roots, pay a dear price when the going gets tough. The tissues of fast-growing plants are packed with protein, nutrients and water, with correspondingly less of tough components such as cellulose and lignin, which inevitably makes them physically weak and a magnet for pathogens and plant-eating animals. This doesn't matter as long as they are able to keep growing quickly, always keeping one step ahead of trouble. But in unproductive environments, the rapid growth and voracious demand for resources that made these plants so successful suddenly become a dangerous liability. Here a quite different strategy pays dividends. Where resources are in short supply, plants need to be less interested in grabbing new supplies than in protecting what they already have. Tough, evergreen leaves, heavily defended against animals, are now the key to success. Cacti, which have abandoned leaves altogether in favour of vicious spines, are one of the pinnacles of this strategy.

The fundamental environmental gradient from good to bad - fertile to infertile conditions; fast-growing to slow-growing species - has dominated the evolution of the world's plants (and animals, for that matter). No single plant can outperform others along more than a tiny section of this gradient. Moreover, it isn't simply one gradient, but many. This is because there's really only one fertile environment, which contains plenty of the goodies that plants need - light, water and nutrients - but myriad infertile ones. A site may be too dry, too wet, too cold, too hot, too shady, too acid, too saline, too unstable, too rocky, or often just too low in nutrients. To paraphrase Tolstoy, all fertile habitats are alike, but infertile habitats vary after their own fashion. At the unproductive extreme of any of these gradients, successful plants will have many crucial differences, depending on the precise details of the environmental difficulties they face, but will have a common characteristic of slow growth and long-lived, well-defended tissues. Nor is there only one way to deal with a particular kind of difficult environment - there are usually at least two, and sometimes more. In seasonal temperate climates, there are several quite different ways to exploit shaded sites beneath deciduous trees. One option is to tolerate low light levels and grow slowly through the shady part of the year. Many ferns adopt this strategy, while wood sorrel is perhaps the arch-exponent among British flowering plants. Another option is to avoid the shade entirely by growing during the spring and early summer, before the tree canopy closes. Among the earliest of these *vernals* are aconites, celandines and wood anemones, while bluebells, ramsons, trilliums and erythroniums emerge a little later. A quite different strategy is to abandon altogether the unequal struggle to obtain enough light, and steal the necessary carbon from the trees. The bizarre parasitic toothwort has no chlorophyll at all and lives on the roots of hazel and elm. The coralroot and bird's-nest orchids have both adopted yet another approach, obtaining their carbon from decaying organic matter, like many fungi.

Drought offers even more possible ways of earning a living. One option, similar to that adopted by the shade-tolerators of woodland, is simply to put up with lack of water. Many desert shrubs do this, sometimes losing their leaves when things get really bad. Another option is to store water from infrequent rains, as do the cacti and many other unrelated desert succulents. Another possibility is to develop an immensely deep root system, able to tap subterranean water supplies even during long periods without rain. Desert annuals don't even try to cope with drought - they spend nearly all their time as buried seeds, emerging briefly to flower, set seed and die only after infrequent heavy rains. When this happens on a grand scale, as in

Namaqualand in southern Africa, it provides one of the world's great botanical spectacles. Finally, as in woodlands, there's always the parasitic option. In Western Australia, the root parasite sandalwood taps water from the roots of its victims, while mistletoe sometimes steals its second-hand water too.

Plants of fertile habitats are less diverse than those of infertile habitats, for reasons I will consider in the next chapter, and high fertility itself constrains how and when such plants grow. All herbaceous plants of fertile habitats have to grow big and fast to overcome their competitors. Inevitably, therefore, they all tend to reach their peak size at the most productive time of year, which in temperate seasonal climates is mid-to-late summer, and then produce flowers on the end of the new growth. They don't have a lot of choice about this sequence, since further growth is largely prevented, or at least dramatically slowed, by the diversion of resources from growth into flowering. This crowding of flowering into midsummer can lead to intense competition for the services of pollinators, so in a few cases natural selection has favoured the strategy of completely separating growth and flowering. A good example is butterbur (*Petasites hybridus*), which produces tall, leafless flowering stems in early spring, followed much later by its enormous summer leaves.

## Evolutionary history and adaptation

The strategy a particular species adopts to cope with drought, shade or anything else is rarely an accident. Plants can only play the hand that their evolutionary history (their *phylogeny*, in biological jargon) has dealt them. Just as you might adopt a quite different approach to crossing a river if you were given either the means of building a boat or a pair of water wings, plants are prisoners of their existing biological inheritance. The penchant of the lily family for growing from bulbs means that they are pre-eminent among the vernal woodland flora throughout the northern hemisphere. Evergreen, slow-growing ferns, common in all kinds of infertile habitats, are ideally qualified to form part of the shade-tolerant woodland ground flora. Sometimes a lack of the necessary variation, particularly in isolated floras, means that a particular solution to a problem is never discovered, in much the same way that the Incas never discovered the wheel. So, while desert succulents are common in America (cacti) and Africa (euphorbias and other families, many of which are startlingly similar to cacti), Australia's desert flora never hit on the idea of genuine stem succulents as one of the answers to drought. Equally, natural selection never wastes anything, so quite often a trait that evolved for one purpose is pressed into service for another. Cacti are familiar features of New World deserts, but characteristics that evolved there are also useful in much wetter habitats. So-called forest cacti grow as epiphytes on tropical trees, where their ability to tolerate drought comes in very handy. Zoology abounds with similar examples, one of the most famous being the use of two reptilian jaw bones to form two of the three bones of the mammalian inner ear (reptile ears make do with one bone).

The converse of the failure of Australian desert plants to invent the succulent habit is that plants can sometimes rediscover something they appeared to have irrevocably lost. One example is the woody habit (the ancestral state in flowering plants), a trait that very much runs in families. Some families, such as oaks or maples, are all woody, while others, such as the cabbage family, are entirely herbaceous. The pea family is one of the few that makes a decent attempt at both. By chance, some small isolated floras (on islands or tropical mountains) started out with few or no trees, forcing plants that are elsewhere entirely herbaceous to reinvent the woody habit. In fact, none of these odd plants can be said to have made a particularly good job of being a tree. The tree daisies of Mount Kenya resemble nothing more than cabbages on giant sticks - but, as Dr Johnson said in another context, the surprising thing is not that it was not done well, but that it was done at all.

## Competition and coexistence

Ecologists tend to assume that plants and animals that live side-by-side can only coexist by having different ways of making a living, by - in ecological jargon - occupying different *niches*. But in fact, and rather surprisingly, even plants that grow next door to each other do not need to be very different. Although the Earth offers many different opportunities for plants to survive, grow and reproduce, and therefore many different *functional types* of plant have evolved to fit these opportunities, there is no requirement for each of these types to have but a single representative.

The biosphere is like a play that offers many varying roles, but does not require that only one actor plays each role. Examination of any diverse plant community could illustrate this principle, although in practice we often don't know enough to understand fully what is going on, especially in diverse tropical communities. One community we do understand, however, is old pasture developed over Chalk and other limestones - arguably the most diverse plant community in northern and central Europe. Such pastures may have 40 or more different flowering plants in a single square metre (plus quite a few mosses). Is it really possible that each of these plants is doing something different, exploiting this small patch of earth in some unique way? It seems unlikely - and indeed attempts to relate the distribution of individual species to variation in, say, soil depth or pH have generally been unsuccessful.[3] Perhaps some pairs or small groups of species have complementary sizes or shapes and fit together like pieces in a jigsaw? Again, probably not - research shows that, in reality, the different species are distributed at random relative to each other. Finally, studies of limestone grassland plants under controlled conditions show that they are genuinely rather similar: nearly all are short, slow-growing, evergreen perennials. The few ways in which these plants divide up their patch of turf are neither obvious nor dramatic. There are subtle and overlapping seasonal waves of growth, with plants at one extreme preferring early spring; others midsummer. There are also profound differences in the requirements for establishment of new plants from seed. Some have big seeds and can germinate and grow in most places; others have tiny seeds and need the big gaps created by rabbits or moles. Seeds of some species live in the soil for decades; others for barely a year. Some germinate exclusively in spring; others only in autumn; some at almost any time of year.

Nevertheless, these differences are not large, and they allow Chalk grassland plants to coexist only because, in this unproductive environment, interactions between the mature plants are slight and rather inconclusive. When growth is slowed by lack of water or nutrients, or plant size is reduced by grazing animals (or both), neighbouring plants may hardly interact at all. Possession may be at least nine points of the law, with little serious opportunity for competition to oust mature individuals, and the main interactions restricted to the brief struggle to inherit the gap when a plant dies or is eaten (and even here, chance may be the main factor). In productive environments things are often very different, and I will say more about how productivity relates to patterns of biodiversity in the next chapter.

If understanding the diversity of Chalk grassland is a challenge, tropical rainforest is a much larger one. Here we have a community with hundreds of different tree species, most of them apparently rather similar, and about most of which we know almost nothing. Tropical diversity is something else I return to in the next chapter.

To sum up: genetic variation, mutation, recombination of genes by sex, and natural selection generate and select biological variability, constantly testing new varieties against the environmental opportunities and challenges available. Isolation, both geographical and biological, allows evolution to proceed relatively independently on every continent, every island and every mountain top, while trade-offs guarantee that no species can be the 'best' in anything other than a very limited range of environments. Finally, even the 'best' solution to a particular environment may have very many different representatives, differing among themselves only subtly or hardly at all. The result is, for example, anything from 300,000 to 400,000 or even more species of higher plants, depending on whom you believe, about 250,000 of which have so far been named. In the next chapter, we will look at where all this plant (and animal) biodiversity is to be found.

Wollemi pine, *Wollemia nobilis*

*Chapter 2*

# Biodiversity: where and why?

*It is, I find, in zoology as it is in botany: all nature is so full, that that district produces the greatest variety which is the most examined.*
**Gilbert White, *The Natural History and Antiquities of Selborne* (1789)**

## Tropical diversity

So where is all this biodiversity, and what gives rise to it? Well, that's not a simple question; in fact it's not even a single question, because the answer is quite different at different scales. At the global scale, most biodiversity is tropical. There are more species of almost everything as you approach the equator, although there are some odd exceptions, like bees and greenfly. For plants the trend is obvious: Britain has about 1,600 native higher plants, yet the whole of the Arctic, north of the tree line (a much larger area), has only a little over 1,000 native higher plants. Antarctica, admittedly with the disadvantage of being almost completely covered by ice and snow, has just two native higher plants. At the other end of the scale, just 14km² of La Selva Forest Reserve in Costa Rica contains more different plants than the whole of Britain. Most other groups show similar trends: Edward O. Wilson famously found a single tree in Panama (and that's an *individual* tree, not a single species) that had more species of ants living in it than in the whole of Britain.

The increase in the diversity of life as we approach the equator is such an obvious pattern that biologists have puzzled over it since at least Darwin's time. Unfortunately, since almost anything that can be measured is different if you compare Manchester with Manila, there is no shortage of possible explanations (more than 25 at the last count). Some of these are obviously true, although not very interesting. For example, cool temperate regions have been repeatedly glaciated over the last few million years, which removed many species that were then unable to return. Britain almost certainly has fewer species than an equivalent area of France, for exactly this reason. Tropical regions very often contain suitable habitat for cold-loving species (on mountains), but of course the converse is not true. However, it seems certain that a large part of the solution is that there's just *more* of the tropics, in terms of both quantity and quality. There's an awful lot of land in the tropics. This is not so obvious from a traditional flat map of the world, owing to the difficulty of fitting a globe on to a flat surface. You could be forgiven for thinking that Brazil and Greenland are about the same size, but in fact the former is four times the area of the latter. Added to that, average temperature hardly varies between 20°N and 20°S, so climatic zones are wider near the equator. Everywhere from Mexico to Paraguay, Mali to Zimbabwe and Burma to Queensland is a single climatic zone. Large area has all kinds of inevitable biological effects. If tropical species have larger ranges on average, they are less prone to extinction and more prone to speciation. Since, at large scales, the number of species is essentially the balance between the appearance of new species and the extinction of old ones, the tropics can hardly help having more species. The tropics are also hotter and therefore, everything else being equal, more productive than temperate and polar regions. More productivity means more biomass (total weight of living organisms), which means the possibility at least of more individual organisms. If all species need a certain minimum number of individuals to sustain a viable population, then more individuals must permit (but not guarantee) more species. More energy could also allow food chains to be longer, which also means more species. You begin to see, I hope, the difficulty in coming up with *the* explanation for tropical diversity.

There may already be at least 25 competing explanations for tropical diversity, but that doesn't stop completely new ones emerging from time to time. A 2006 comparison of plants from the lowland tropics and from tropical mountains[1] found that the former had twice the mutation rate of the latter, and much the same thing has also been found in hummingbirds from these two habitats. It seems that the chance of one DNA nucleotide being replaced by another increases with metabolic rate, which itself increases with temperature. Not only that, but tropical organisms also generally manage to get through more generations per year than temperate ones. In other words, even if everything else is equal, heritable variation (the raw material of evolution) just crops up faster in the tropics.

## Biodiversity hotspots

Still at the global scale, but focusing in a little, we begin to see that diversity is very unevenly distributed indeed across the planet. If we look at endemic plants (those that grow in a single region and nowhere else), we can identify biodiversity 'hotspots' that each contain at least 0.5 per cent of the world's 300,000 higher plant species as endemics.[2] There are 25 of these regions, which together contain nearly half of all the world's plant species. Before widespread habitat destruction by humans (of which more in Chapter 5), they occupied 11.8 per cent of the Earth's land surface, but they have lost on average 88 per cent of their original natural vegetation, so they now occupy just 2.1 million km², or 1.4 per cent of the Earth's land. Sixteen hotspots are tropical and nine consist entirely or mainly of islands – indeed nearly all tropical islands are in one hotspot or another. The next largest category after the tropics is Mediterranean climates; all five of the world's Mediterranean climate zones (the Mediterranean basin itself, California, Chile, the Cape region of South Africa and south-west Australia) are hotspots.

The sheer richness of some of these hotspots is bewildering to a botanist raised in the comparatively boring temperate zone of Europe. Take the Cape Floristic province: a tiny piece of South Africa about the same size as Ireland. The latter has about 500 native higher plants; the Cape has 9,000, over 5,000 of them found nowhere else. The dominant shrubby members of the Proteaceae give rise to the characteristic vegetation: *fynbos*, or fine-leaved bush. There are the weird, rush-like restioids, over 800 species of *Erica*, and the largest collection of bulbs anywhere in the world: 1,400 species in all, including 96 species of *Gladiolus* alone. And this box of delights thrives in the most unforgiving and hostile of habitats – frequent devastating fires and soils so poor and acid that the early settlers wrote them off as completely useless for agriculture.

Why are these areas so rich in endemic plants (and, incidentally, almost equally rich in endemic vertebrates)? There may not be a single, simple explanation, but at least a part of it is topography: few hotspots are flat. The most diverse vegetation in the world is on the border between Colombia and Ecuador, where lowland tropical diversity meets the extraordinary topographic complexity of the mountains of the Cordillera Real and the Cordillera Occidental. Writing late in his life, the Prussian explorer and naturalist Alexander von Humboldt already recognised that this area, which he explored in 1801-02, must be a candidate for the ultimate biodiversity hotspot:

> *This portion of the surface of the globe affords in the smallest space the greatest possible variety of impressions from the contemplation of nature. . . . There, the different climates are ranged the one above the other, stage by stage, like the vegetable zones, whose succession they limit; and there the observer may readily trace the laws that regulate the diminution of heat, as they stand indelibly inscribed on the rocky walls and abrupt declivities of the Cordilleras.*[3]

Topography is a major cause (perhaps *the* major cause) of local environmental variation. Apart from the climatic variation associated with altitude itself, interesting topography is associated with diversity in soil depth, water availability and aspect. In this respect, topography is just one example of the role of spatial heterogeneity in generating biological diversity. Indeed, so important is spatial heterogeneity that it can completely override the otherwise universal 'rule' that larger areas contain larger numbers of species. In one set of nature reserves in central Hungary, varying in size from less than 100ha to more than 10,000ha, the smallest reserves supported the same number of insect species as the largest, simply because they contained a greater variety of habitats. Heterogeneity is so crucial that it can sometimes even overcome the effect of area *and* the ubiquitous tropical-temperate gradient of biodiversity: a large area of very flat topography in tropical west Africa has only 20 per cent more species than Spain and Portugal, despite being eight times larger. Often heterogeneity is obvious, but it doesn't have to be; mudflats may all look the same to you and me, but variation in particle sizes from place to place generates different animal communities.[4]

The extraordinarily diverse Cape region is mostly acid sands, but there are pockets of limestone; ranges of hills and mountains from sea level to over 1,000 metres, hot in the lower parts but with snow on the mountain tops; rainfall from as little as 400mm up to 1500mm per year. This isn't enough to fully account for its biodiversity, of course – plenty of places have the Cape's physical assets but lack its biological riches. In evolutionary terms, the Cape's plants are astonishingly young, which perhaps explains how many of them manage to be so rare – there are only a few hundred individuals of many Proteaceae. Are these future successes at the start of their careers, or failed evolutionary experiments on their way to extinction? We still don't know.

Of course, the world being the surprising place it is, just when you think you can see the wood for the trees, you come across something that ruins your convenient explanation. While varied topography is undoubtedly important for biodiversity, it doesn't work everywhere. In a tiny corner of south-western Australia we find *kwongan*, Australia's answer to the fynbos of South Africa. With a similar climate to the fynbos and even poorer soils (often more-or-less pure sand), kwongan heathlands look dull but are almost unbelievably diverse. A patch of kwongan only 10m by 10m may contain 110 different species of plants. A similar patch a kilometre away will also contain 110 species, but only half of them the same as in the first patch. Altogether there are around 7,000 species of plants in this area, 85 per cent of them found nowhere else in the world. The lifestyles of many of these plants show just how hard it is to make a living here: an astonishing diversity of parasites (including the world's largest mistletoe, a tree up to 10m tall) and the world's greatest concentration of carnivorous sundews. Nor is the diversity confined to plants – Western Australia is also home to 141 different mammals and 439 reptiles, of which 25 and 187 respectively are unique to WA. And – you guessed it – the kwongan is almost flat; the monotony relieved only by a few low, flat-topped hills.

# Plants and soil pH

Some global patterns of plant diversity look simple at first, but on closer inspection turn out to be quite baffling. If, like me, you learned your botany in Europe, you soon realised that there's a very close connection between plant diversity and soil pH. High-pH (*basic*) soils on Chalk and other limestones support rich floras, while low-pH (*acid*) soils are dull. The patch of British grassland that allegedly inspired David Bellamy's lifelong passion for plants (now helpfully labelled *Bellamy's Bank* on the obligatory interpretation panel), in Miller's Dale, Derbyshire, is on limestone and has a neutral soil pH, of around 7. Old, unfertilised limestone pastures like Bellamy's Bank often contain up to 40 species per square metre, and average around 25 species. Acid grassland (with a pH around 4) averages only around seven species per square metre, and hardly ever contains over 15 per square metre. To put these numbers in perspective, the typical suburban British lawn might contain around 25 species in all, while a square metre would contain on average around ten or a dozen species. (It might not look like that many, but many of them are grasses which really do all look the same, especially when regularly mown.) The diversity of small samples of pasture or meadow on soils of different pH is a faithful reflection of the numbers of plants suited to that pH in the landscape as a whole: however many British grasslands of pH below 4 you look at, you will never find more than about 60 species in total, but grasslands of pH around 7 support about three times as many. In other words, there aren't just more species per square metre on Chalk and limestone, there are also more species altogether.

After a while, it's easy to convince yourself that some kind of natural law must make the floras of basic soils interesting and those of acid soils boring. Therefore one of the most puzzling discoveries European botanists made, as they fanned out over the globe, must have been the frequent tendency of the pH–diversity relationship to go into reverse. We have already seen that two of the most diverse kinds of vegetation in the world, South African fynbos and Australian kwongan, occur largely on acid sands, and this is true also of the hyper-diverse Brazilian savannah known as Cerrado. So what on Earth is going on here? The truth is that there is no good reason for either high-pH *or* low-pH soils to support diverse floras, and the relationship in any one spot may simply be a historical accident. Soil conditions that exist over large areas and for long periods eventually give rise, by the slow workings of evolution, to large floras. Europe, and particularly the Mediterranean, the cradle of the European flora, has always contained large areas of limestone, so most European plants are adapted to limestone soils. Elsewhere, there have often been large, persistent areas of acid soil, so most plants have evolved to exploit acid soils. Sometimes acid soils were a direct consequence of underlying calcium-poor, sandstone rocks, but acid soils are common in the tropics for another reason. High temperatures, often combined with high rainfall, soon break down carbonates and leach all the calcium from tropical soils (calcium carbonate - or lime - is the main reason for high-pH soils). There are exceptions, but acid soils are the rule in the tropics. In fact, the irony for limestone-fixated European botanists like me is that over the whole globe there are many more plants of acid soils than there are of basic ones. A graphic demonstration of the fickle relationship between soil pH and diversity comes from studies of remnants of diverse grasslands in south-eastern Australia. These grassland fragments are heavily invaded by alien plants from Europe, but if you compare individual small patches, the diversity of the native Australian plants goes *up* as the pH goes *down*, while the diversity of the European invaders does exactly the opposite.

A postscript to the soil-pH story is that it applies only to relatively 'moderate' pH, say between 4 and 9. At extremely acid pH, aluminium and manganese become soluble in toxic quantities, while essential nutrients (especially phosphorus) are unavailable at *both* extremes. Because survival of plants of any sort is low at extreme pH, plant diversity at extreme high and low pH is low, irrespective of the arguments made in the previous paragraph. On Bleaklow, an acid upland between Manchester and Sheffield in northern England, a combination of soils that were acid to begin with and two centuries of industrial acid rain has led to soil pH values approaching 2. At that pH not even the local acid-tolerant flora will grow unless the pH is raised by liming.

# Success at small scales

So it is generally accepted that, at the global or regional scale, biodiversity is a product of evolution. Even if we don't always understand how it happened, we know that the Cape flora is rich because many species have evolved to live there, and the flora of acid soils in Europe is poor because few species evolved there. However, as we focus down to a scale of a few metres, or tens of metres, the size of the pool of species supplied by evolution often becomes less and less important. Here, diversity is often controlled by something we considered earlier at the global scale, but in quite a different way: productivity - or favourability for plant growth. Productivity has a profound impact on small-scale plant diversity. To understand why, it helps to visualise productivity, in a general sense, as a wheel, or better still as a dartboard, which is fertile at the centre and infertile at the edges. At the hub or bullseye, there are plenty of all the things a plant needs (nutrients, water, light, carbon dioxide), and not too much of anything nasty, like salt or heavy metals (or even water - few plants tolerate waterlogging). As we move away from the bullseye in different directions, things become gradually worse in different ways. In one direction it gets too shady; in another too dry; in another the soil becomes too shallow and stony. Beyond the edge of the board, things are so bad that there are no plants at all, or at least no higher plants - lichens and some mosses carry on into territory well beyond the tolerance of higher plants.

The exact identities of the plants on the different parts of the dartboard vary according to climate and other factors. For example, in wetlands in North America or Europe, the bullseye (undisturbed sites with deep, fertile soils) is often monopolised by a species of *Typha* (cattail, reedmace or bulrush), up to 2m tall and spreading below ground to form a mat of rhizomes. As we move away from the bullseye in different directions, the plants that come and go are dissimilar, depending on exactly what prevents *Typha* from dominating. As I pointed out in Chapter 1, there are many possibilities: for example, shallow gravelly or sandy soils, erosion by wave action, rivers or animals, or occasional drought. In every case we tend to move through a zone of large sedges, then smaller sedges, then a mixture of sundews, grasses, annuals and other plants, and finally no higher plants at all. This pattern gives a clue to one reason for the relative scarcity of species adapted to fertile, undisturbed conditions. Because such habitats are generally dominated by a small number of superlative competitors (in the extreme case, just one), there has been little opportunity for the evolution of a wide diversity of potential inhabitants. A second factor is analogous to the effect of soil pH, described on pages 34-6. Most of the globe, throughout most of geological time, has been relatively infertile, owing to low temperatures or lack of water or nutrients. The activities of modern society, including irrigation and the application of fertilisers, tend to obscure the fact that until very recently opportunities for the evolution of plants of fertile habitats have been few and far between. Before we started manufacturing large quantities of artificial fertilisers, high soil fertility tended to be confined to floodplains and especially estuaries, where reliable moisture coincided with nutrients washed down from surrounding plains and mountains. It's no accident that agriculture and indeed Western civilisation itself first arose in the 'fertile crescent' comprising the basins of the rivers Nile, Tigris and Euphrates.

## Small-scale diversity

So, very different kinds of plant grow in different places on the dartboard, but where are they most diverse? Where do we find the greatest number of plant species growing together in the same few square metres? Not at the centre, because here the ideal conditions lead to the dominance of the biggest, best and fastest-growing plant available in the local flora. In northern Europe this might have been bramble, bracken, nettles or hogweed, but now it's quite likely to be Japanese knotweed. These are plants that take no prisoners, monopolise all the light and nutrients available, and often exist as natural monocultures. Nor is diversity high at the edge of the board either, because here the existence of plants of any sort is a precarious business, and very few have evolved to exploit such unprofitable conditions. At the very edge of the board, just before life becomes too tough for any higher plant, there is often just a single species. So, if diversity is usually low at the extremes (although for very different reasons), diverse vegetation must logically be somewhere in the middle. In fact, both theory and observations suggest that the most diverse vegetation is found relatively near the edge of the board, where things are moderately (but not *very*) poor for plant growth. The reason is simple - the most diverse vegetation is found where things are good enough for a wide range of plants to survive, but poor enough to prevent any single plant from excluding most of its competitors. It doesn't take much surplus fertility for this exclusion to begin to happen and, however different all the world's really diverse vegetation types may appear to be, all of them (from fynbos to tropical rainforest, via Cerrado and Chalk grassland) occur on soils that are low in phosphorus. As we will see in Chapter 5, an excess of nutrients is responsible for much of the trend of declining biodiversity in northern Europe.

Sometimes, we know enough to define quite closely where the 'window of opportunity' for diverse vegetation, between the low diversity at high and low fertility, actually occurs. For example, in the semi-natural herbaceous vegetation of northern Europe, the window can be defined in terms of the maximum weight of living (and recently dead) biomass. 'Weight' here means dry weight, because water is both heavy and present in plants in rather variable amounts, making wet weight a very unreliable quantity indeed. If dry weight is between 350 and 750 grams per square metre, then diverse vegetation is possible, although by no means guaranteed (for example, diversity may still be low on very acid soils). However, we can say with some certainty that species-rich vegetation does not occur outside this range. In northern England, as we move away from this potentially diverse region towards higher biomass, we encounter low-diversity vegetation dominated by meadowsweet and butterbur (about 900g per square metre), nettles (1-1.5kg), rosebay willowherb (about 2kg) and bracken (2.5kg). Bracken weighs in at the top because of its secret weapon - a dense blanket of dead fronds that swamps all potential competitors even before the bracken starts to grow in early summer. But it now has competitors in Japanese knotweed and giant hogweed that can - in the right habitat - do even better.

Sometimes, low fertility itself is the only thing stopping bigger, faster-growing plants from taking over. More often than not, however, there is another important factor: management, usually in the form of cutting or grazing. Grazing is very good at preventing the exclusion of small plants by more vigorous competitors, but is not enough on its own; grazing can produce really diverse vegetation only when combined with low-to-moderate fertility. It's not surprising that limestone pastures are the most diverse vegetation in northern Europe, since they combine all the right environmental variables in the same place: grazing, high-pH soils and low fertility.

In a generally fertile landscape, such as much of the lowlands of northern and central Europe, there may be little vegetation below the lower biomass threshold of 350 grams per square metre. What there is may be confined to deep shade, sand dunes and rock outcrops, or some very infertile man-made habitats such as mine spoil, cinders and brick rubble. Because of this, the 'hump-backed' relationship between fertility and diversity (diversity low at both ends of the fertility gradient and high in the middle) may not be obvious, and it may look more like a simple negative relationship: higher fertility = lower diversity.

A massive study[5] in 1998 of 281 European grasslands in Belgium, The Netherlands, UK, Spain and Luxembourg illustrates this relationship between fertility and diversity perfectly. All were old grasslands, in which species composition and diversity can reasonably be assumed to be in equilibrium with the physical environment, and all were managed by grazing or cutting. The diversity of these grasslands turned out to be very strongly dependent on soil fertility, and in particular on the amount of phosphorus in the soil. Although phosphorus varied from almost zero to 35 milligrams per 100 grams of soil, diverse vegetation was absolutely confined to soils with less than 5mg of phosphorus. Below this value, the grasslands contained anything up to 60 different species per 100 square metres. Above it, no grassland contained more than 20 species. Moreover, rare plants were completely absent from grasslands above the 5mg threshold.

Whether we measure favourability for plant growth by the biomass actually achieved (as in the UK example) or by soil nutrients (as in the European example), the range compatible with high diversity has a very striking and rather sharp upper threshold. Between 750 and 1,000 grams of dry matter, or between 5 and 6 milligrams of phosphorus, diversity falls over a cliff. Over most of the possible range of fertility (from the bullseye to quite near the edge of the dartboard described above), plant diversity is low. If you want to preserve species-rich vegetation, it seems that there is no really 'safe' dose of phosphorus fertiliser.

# Tropical trees

These principles of biodiversity levels at the small scale apply everywhere, although the numbers are not always precisely the same. For example, there is some evidence that the 'window' of high diversity is in a slightly different place in tropical grasslands, although again the overall shape of the relationship between biomass and diversity is the same. However, sometimes we encounter biodiversity of such awe-inspiring magnitude that it's clear there has to be something else going on. One such case is tropical trees. One square kilometre of rainforest in Ecuador or Borneo may contain more different kinds of tree than the entire northern hemisphere outside the tropics - an area over four million times larger. From that simple fact we can deduce an obvious corollary: that almost all tropical trees are (by temperate standards) desperately rare. This prompts all kinds of question, but the key one is: What stops one or a few tropical trees becoming common and excluding most of the others? In other words, what stops the diversity of a tropical forest resembling that of a typical temperate one, dominated by oak, beech, maple or ash?

We do not yet have a complete answer to that question, and indeed the topic remains one of fierce debate between ecologists. Just as with high tropical diversity in general, there may not be a single answer, but the front-runner remains an idea suggested independently 40 years ago by two Americans, Dan Janzen and Joseph Connell.[6,7] Janzen and Connell's theory brings together several notions: (a) in the constant, benign climate, tropical trees suffer much more from herbivory and disease than temperate ones; (b) tropical herbivores and pathogens tend to be more host-specific than temperate ones; and (c) seeds and seedlings of tropical trees suffer much more damage if they occur in dense patches and/or if they are near the parent tree, which tends to act as a source of specialist herbivores and pathogens. In other words, the safest place for a tropical tree seedling is as far as possible from its siblings and (especially) from its parents. It's easy to see that the Janzen-Connell theory prevents individuals of any tropical tree growing too close together, which effectively prevents any one species becoming common. Numerous experimental tests have produced strong evidence in support of the theory, and the debate has now shifted from whether it's true to whether it's *enough* - a question on which the jury is still out.

# The key role of dispersal

So high plant diversity is possible wherever: (a) the environment is poor enough to prevent a few good competitors from taking over, but

(b) it is not so bad that hardly anything can grow at all, and (c) natural selection has provided plenty of suitable species. Oh, and (d) there are probably a few extra things going on in the tropics, even if we still aren't sure what they are.

Is that it? Well, not quite. In Britain, a square metre of a diverse limestone grassland may contain around 35 different flowering plants. Two neighbouring square metres might have 40 plants, and four square metres a few more, but we're still a long way from the 200 or so species in the British flora that we know *could* grow there. Inspection will reveal that many of the missing species simply aren't in the local area, illustrating the important point that local diversity is nearly always limited by the failure of some species to disperse to the site at all. Sometimes this apparent dispersal failure is simply a matter of time. The island of Krakatoa was completely sterilised by the cataclysmic volcanic explosion of August 1883 and, although both plants and animals began to recolonise almost immediately, the number of species on the island is still increasing. The better dispersers – those that could fly or float – returned quickly, but poorer dispersers take longer. Many suitable species may never make it across the sea from Java or Sumatra.

Woodland plants are notoriously poor dispersers. One study[8] in Pennsylvania looked at the spread of woodland understorey plants from old woodland (containing many specialist woodland plants) to new woodland derived from abandoned agricultural land (containing hardly any woodland plants). Even though the two woodlands formed one continuous block, even the very best-dispersed woodland specialists were spreading into the new woodland at a rate of only about two metres per year, and most had achieved less than half that rate. Even 50 years after canopy closure in the new woodland, five species hadn't made any progress at all and were still firmly stuck in their original habitat. In Britain there is a long list of 'indicator species' of ancient woodland. These are plants confined to woodland that has never been felled. They could grow in many newer woodlands – they just can't get there.

The obvious conclusion – that most plant (and animal) communities contain fewer species than *could* live there – is simple to test. It's easy enough to sow seeds of new, apparently-suitable-but-not-actually-present plants into existing communities. The usual finding is that yes, local diversity can easily be increased by this simple expedient. In ecological jargon, few communities are 'saturated' with species. Low diversity due to dispersal failure is common and has two important practical consequences for conservation.

First, there is a burgeoning modern science of ecological restoration, which attempts to restore habitats that have been damaged - by mining, drainage, pollution or agriculture - to something like their former state. Nearly always, however perfect the restoration of the physical environment, much of the desired biodiversity fails to show up. Because many plants and animals are poorly dispersed, this can happen even if diverse, pristine habitat is nearby. In the modern landscape, however, there is often no undisturbed habitat in the vicinity, which only makes the problem worse. In fact, fragmentation of natural habitats by farms, towns and roads is itself a major cause of local species extinctions, even if the surviving patches of habitat still look healthy. We'll have another look at the effects of fragmentation in Chapter 5. In the case of plants, the problem can often be solved by sowing seeds of the 'missing' species. Of course, establishment from seed is inherently risky, and although increasing local diversity by sowing seeds is cheap, it's often safer to grow plants in pots and then plant them out. (One American ecologist explained to me that this is *always* a good idea, however certain you are that establishment from seed will work in the end. "Really?" I said. "Why?" "Because," he replied, "when the senator who voted the funds for the project turns up, he wants his photo taken with some concrete sign that it worked.")

Second, there is a natural tendency to assume that the plants and animals we find living in a particular spot are peculiarly (or even uniquely) well-adapted to live there. This is the prevailing dogma among a certain kind of patriotic, native-is-best type of conservationist, but it's not true. The species that live anywhere are merely the best (or perhaps just the luckiest) among the fortunate few that actually succeeded in reaching that particular spot. Some ecologists spend their lives trying to explain why so many plants and animals are able to do so extraordinarily well (and sometimes cause so much trouble) when transplanted to new countries, or to new continents. In fact there is nothing to explain, bearing in mind the vanishingly low probability - given the whole world's biota to choose from - that the present incumbents anywhere are *the* best-adapted for that location.

## Diversity in action: the Park Grass plots

To see in a single location all the forces that determine local diversity in action, we need look no further than the world's oldest ecological experiment: the Park Grass plots at Rothamsted, Hertfordshire, England.[9] The Park Grass Experiment was begun in 1856 by Sir John Lawes, one of the pioneers of the study of fertilisers on plant growth. Lawes took a field that had been pasture already for at least a century and laid out a range of fertiliser treatments that continue with minor modifications to this day.

Some plots have received high rates of all the major nutrients (nitrogen, phosphorus and potassium) that plants need to grow well, and these plots have low diversity (around 10-12 species), owing to dominance by relatively few good competitors, chiefly the large tussock grasses *Arrhenatherum elatius*, *Dactylis glomerata* and *Alopecurus pratensis* (note that diversity never gets *very* low because the grassland is maintained by annual cutting, which helps to control these potential dominants). The site started out with a relatively high soil pH of about 6, but some plots have received a combination of treatments (high rates of ammonium sulphate and no lime) that have reduced soil pH to remarkably low values. The most acid plot (pH 3.6) now contains only two species, the grasses Yorkshire fog (*Holcus lanatus*) and sweet vernal grass (*Anthoxanthum odoratum*), which are the only species in the neighbourhood that can tolerate such an extreme pH. Such extremely low diversity is a first-rate example of local diversity limited both by extreme pH and by dispersal. There *are* other species that could grow at pH 3.6, but they are mostly found in acid uplands far to the north and west of Rothamsted and have no hope of reaching the site. The nearest bilberry to Rothamsted, for example, is on the Greensand Ridge at least 30km away, and even that is an isolated and rather forlorn outlier.

Finally, the most diverse plots (over 40 species) have received no fertiliser at all and are in that low-fertility, high-pH 'window' that allows plenty of species to survive but prevents any of them from taking over. Despite their high diversity, however, these plots also illustrate the limitation of diversity by dispersal. They now contain virtually *all* the suitable species that are available locally and, given the domination of the local landscape by intensive arable farming, there is now little chance of any further species turning up. We can say with some confidence, however, that were we permitted to tinker with the experiment by sowing more species (which we certainly aren't!), local diversity could easily be increased.

Now we know a bit more about diversity, we'll soon be able to tackle the big questions - such as what biodiversity is worth, why it is threatened, and what we can do about it. First, however, we need to look deeper into species.

Greater reedmace, *Typha latifolia*

*Chapter 3*

# Inside species

*When you have seen one ant, one bird, one tree, you have not seen them all.*

Edward O. Wilson, *Time* magazine, 13 Oct 1986

## There are more bats - and everything else - than we thought

Britain has 15 breeding bat species, or at least that's what everyone used to think. Bats fly at night and are hard to see, so they are often detected and their populations monitored by listening - electronically - to their high-pitched echolocation calls. In the early 1990s it began to dawn on a few people that not all individuals of Britain's commonest bat, the pipistrelle, sounded the same. Some, in fact most, seemed to call with a frequency that peaked at around 46kHz, but the calls of others peaked at a higher frequency, around 55kHz. Further enquiry revealed that the two types were clearly separate and, although both eat small insects such as flies and midges, the 46kHz type is a generalist of deciduous woodland while the 55kHz bat always forages close to water. Finally, analyses of the DNA of the two bats revealed that they are quite distinct species that separated five or perhaps even ten million years ago. In other words, they are much more different than, say, modern humans and Neanderthals, which split less than one million years ago.

So Britain now has two pipistrelles; the original (lower-pitched) common pipistrelle, and an extra species known - by virtue of its higher pitched call - as the soprano pipistrelle. Now we know what to look for, it seems astonishing that we ever thought they were a single species: they don't interbreed, and their calls, foraging and roosting behaviour turn out to be quite distinct, so why did it take so long to tell them apart? The problem is that, traditionally, species have been described and separated largely on the basis of their appearance, or *morphology*, and the two pipistrelles look so extremely similar that even experts find it hard to separate them. But humans are very visually oriented animals, while many animals rely much more on sound, behaviour and scent. The two pipistrelles, in fact, are a classic example of what are now called 'cryptic species', although of course they are only cryptic to *us* - they have no trouble at all telling each other apart.

The fact that the double life of Britain's commonest bat was discovered only recently, in a country with few bats and very many (professional and amateur) naturalists, might lead you to suspect that there are many more cryptic species out there than we know about. That suspicion would be correct. To start with, the discovery of cryptic species was largely accidental. For example, one ecologist, who had spent years studying the pollination biology of tropical vines in the genus *Dalechampia*, wanted to cross-pollinate populations of *D. scandens* from five widely separated locations in Mexico. When he tried, however, he found that they wouldn't: although they looked more or less identical, they were different species. More recently, progress has accelerated with the advent of 'DNA barcoding', in which a short, standardised DNA region is used to identify species. For technical reasons the DNA chosen is usually from mitochondria (in animals) or plastids (in plants), intracellular organelles with their own complement of DNA, independent of that in the nucleus. DNA barcoding has allowed the systematic search for cryptic species to begin, and, in a striking illustration of how hard it is to find something until you know it's there, they now turn up everywhere. To pick one example at random from the recent literature, barcoding 16 apparently 'generalist' parasitic flies (i.e. attacking many different hosts) from Costa Rica has revealed 73 species, a few of them generalists but the great majority highly specialised.

Large areas of ecology, as well as our understanding of the true extent of the Earth's biodiversity, are in the process of being turned upside down by these discoveries. As in the example above, many insect herbivores, predators and parasites that were previously thought to be dietary generalists are being revealed as complexes of dietary specialists. Conversely, some symbiotic interactions seem to be less specialised than we believed. The flowers of trees in the large tropical fig genus are famously pollinated by tiny fig wasps, and any textbook will tell you that each fig has its own, single fig wasp, but we now know that at least half of all fig species are pollinated by more than one (cryptic) wasp species.

Nor are cryptic species just an academic curiosity. Cryptic species have shown up among pathogens of both humans and crop plants, with important implications for our attempts to control such diseases. The mosquito *Anopheles gambiae*, the primary carrier of human malaria in Africa, is actually seven cryptic species that vary in habitat and host preferences. Some of the new species don't attack humans at all, which implies that at least some control efforts have been misguided. A failure to identify cryptic species of venomous snakes might even cost you your life if you are bitten by one, since we know that closely related species often have very different venom - cryptic species have been identified among pit vipers and Asian cobras. Even the common and commercially valuable blue mussel, *Mytilus edulis*, turns out to be a complex of three cryptic species. Mussels are commonly used to monitor pollution, but the three species have different growth rates, leading to inaccuracies that could endanger human health.

Conservation of threatened species will inevitably be complicated by increasing recognition of cryptic species. Some endangered species must consist of groups of cryptic species, each of which is even more endangered than we thought, and (worse still) might require different conservation strategies. For example, research in 2010[1] has confirmed that the critically endangered Kemps ridley marine turtle (*Lepidochelys kempii*) really is distinct from the widely distributed and very similar olive ridley turtle (*Lepidochelys olivacea*). Barcoding plants has lagged somewhat behind barcoding animals, but recommendations are now emerging for a gene that would serve as a universal barcode for seed plants at least. The practical implications are huge. Barcoding could enormously speed up our ability to produce biodiversity inventories of diversity hotspots, while on a more immediately practical level, customs officers could use barcodes to identify plants seized while being traded illegally. For example, all orchids are in Appendix 2 of the Convention on International Trade in Endangered Species

(CITES), which means a permit is required for their trade, but the proposed barcode easily identifies those orchids (e.g. the Central American *Phragmipedium*) that are in Appendix 1, for which all trade is prohibited. Last but not least, cryptic species are beginning to make us think again about the question with which this book began: How many species are there? Critics claimed that Terry Erwin's estimate of anything up to 30 million tropical arthropods was based on an inflated estimate of dietary specialisation, but recognition of cryptic species is now starting to make that total look conservative.

# Genetic rescue

Of course, every species, cryptic or otherwise, consists of individuals that differ genetically from each other in small ways, and we often appreciate this genetic diversity only when we start to lose it. One of the consequences of habitat destruction is that populations of some animals may be reduced to very few individuals with very low genetic diversity, leading to inbreeding. This is particularly likely to happen with large animals that only ever occurred at low densities, such as large predators. The cougar (or mountain lion) used to range throughout the New World but, except for the isolated Florida panther (*Puma concolor coryi*), it has been exterminated in eastern North America. Even there it has come perilously close to extinction, down to around 30 animals in the 1980s. But even that was probably not the low point: scientists who looked at the DNA of the surviving animals found almost no genetic variation between individuals, and suggested that at some point (probably in the 1960s) there may have been only a single pair, from which all surviving animals are descended. Not surprisingly, such extreme inbreeding led to inherited deformities, including low sperm count and undescended testicles.

The plight of the Florida panther led to a vigorous argument about whether a 'genetic rescue', i.e. importing cats from other areas, should be attempted. The arguments for and against were long and complex, but the final piece in the jigsaw was confirmation from the Fish and Wildlife Service that the Florida panther would still deserve its protection as a Federally listed endangered species even if the population contained genes from other subspecies. The rescue plan finally went ahead in 1995, with the introduction of eight cats of the Texan subspecies, *P. concolor stanleyana,* and the results have thoroughly vindicated the optimists. 'Hybrid' kittens are three times more likely to reach adulthood than pure Florida kittens, and the population has risen to a recent count of 87. Moreover, the cats have now spread into areas that some argued (before the rescue) were unsuitable. The Florida panther isn't safe, but it's a lot safer than it was, even if it isn't quite a Florida panther any more.[2]

## Losing genes

Later, in Chapter 5, we'll take a look at threats to biodiversity, and that discussion will, for entirely pragmatic reasons, be entirely about loss of *species*. But of course extinction of an entire species is merely the end point of an incremental process of extinction of individual populations. As a species loses its component populations it also slowly loses its ability to adapt, its role in the ecosystem, and ultimately (as in the case of the Florida panther) its ability to survive at all. So what can we say about the diversity of the component populations of species and their rate of extinction?

First of all we need to define what we mean by a population. In this context, the genetic or Mendelian (after Gregor Mendel, who first identified genes as the units of inheritance) definition is most appropriate: a Mendelian population is a group of individuals that form a more-or-less independent evolutionary unit, owing to limited gene flow with other, genetically distinct populations. Sparing you (and me) the mathematical details, it's possible to calculate that individual species occupy (on average) around 2.2 million $km^2$, while their component populations occupy (on average) $10,000km^2$. Simple arithmetic therefore says there are 220 populations per species on average, which means (if the higher estimates of global species numbers are right) around 5 billion (or more) separate populations. For various reasons, this is certainly a conservative estimate.

Even getting this far involves some big assumptions, and deciding the rate at which we are losing populations - and the genetic diversity they contain - involves a few more. In Chapter 5 we will see how difficult it is to estimate extinction rates of species, even in the comparatively few groups of organisms we know a lot about. However, if we take a fairly middling view of how many species there are, and of the rate at which they are going extinct, we arrive at an educated guess of about 16 million populations lost per year in tropical forests alone, or about 1,800 per hour. Should we be worried about this? Or, since we certainly should be worried, *how much* should we be worried? Before we can answer that question, we need to ask what we get from biodiversity, which is the subject of the next chapter.

Common pipistrelle bat, *Pipistrellus pipistrellus*

*Chapter 4*

# What is biodiversity worth?

*It seems to me that the natural world is the greatest source of excitement, the greatest source of visual beauty, the greatest source of intellectual interest. It is the greatest source of so much in life that makes life worth living.*

**Sir David Attenborough, BBC interview, 2006**

## Crops, medicines and beauty

Rice, *Oryza sativa*, is one of the world's six great staple crops. During the 1960s, new high-yielding varieties of rice were developed by the International Rice Research Institute (IRRI) in the Philippines. These new varieties provided ideal conditions for the evolution of new strains of virus diseases, and rice throughout Southeast Asia was soon infested by a new strain of grassy stunt virus, spread by the brown planthopper. Rice plants infected with grassy stunt have large numbers of small shoots with short, stiff, narrow leaves. The leaves themselves are yellow-green with rusty spots. Plants that become infected at an early stage of growth have low yields, and since millions of people depend on rice, disaster loomed. Fortunately, over six thousand rice species and varieties from all over the world are stored at IRRI, and a search of these revealed just one collection of a wild rice relative, *Oryza nivara*, that was resistant to grassy stunt. The resistance gene was rapidly incorporated into the new varieties, which proved very successful after their release in 1976. Of course, the virus didn't take this lying down, and new mutated varieties soon began to find ways round the resistance. By exploiting the huge genetic diversity at their disposal, however, rice breeders have so far managed to stay one step ahead of the virus. Ironically, excessive use of pesticides almost certainly caused the problem in the first place, by wiping out the natural predators of the brown planthopper.

In fact, even the most hard-headed economists agree that conserving crop diversity in gene banks makes good economic sense. This is because although the facilities are relatively expensive to construct, they can then store very large numbers of plant collections. The extra or 'marginal' cost of protecting each extra collection may therefore be very small. For example, the International Centre for Maize and Wheat Improvement in Mexico holds 17,000 maize collections and 123,000 wheat collections. The Millennium Seed Bank in the UK (part of Kew Gardens) already stores seeds of the entire British flora and aims eventually to have seeds of 10 per cent of the world's flora under lock and key in a bomb-proof freezer in southern England. The value of such collections for crop improvement has been immeasurably advanced by the advent of modern biotechnology. No longer do those who want to improve the disease resistance of rice have to rely on rice alone for their raw material. Genes from unrelated plants can now be used to confer disease resistance on crop species, although this technology is not without its opponents. Interestingly, the boundary between modern 'genetic manipulation' and traditional plant breeding is not as clear as some people think, and the exploitation of the virus-resistance gene in rice could not have been achieved without modern 'embryo rescue' technology. Crosses between different species are often difficult to obtain by conventional methods. Fertilisation may not take place and, even where it does, the hybrid embryo may abort before it develops into a mature seed. One way round this is to 'rescue' the embryo at an early stage and culture it outside the plant, which is how the grassy-stunt-resistance gene was introduced into commercial rice varieties. *Oryza nivara* itself is agronomically worthless and was valuable only because the resistance gene could be transferred to crop varieties.

Crop breeding, however, is far from being the only significant economic reward for biodiversity conservation. Naturally occurring plant chemicals form the basis of a large part of the world's pharmaceutical industry. An oft-quoted example is the rosy periwinkle (*Catharanthus rosea*) from Madagascar, which contains two alkaloids with remarkable cancer-fighting activity. Vinblastine is very effective against Hodgkin's disease, while vincristine is almost a complete cure for childhood leukaemia. The Eli Lilly Corporation in the USA has reputedly earned over $200 million per annum from sales of these two compounds.

Many other important drugs continue to be extracted from plants, since the alternatives of chemical synthesis or genetic engineering are rarely economically viable. For example, even today the anti-malarial alkaloid quinine is still extracted in large quantities from the bark of *Cinchona* trees.

There is every reason to believe that many more drugs await discovery. Few of the plant medicines known to indigenous people have been scientifically investigated. For example, recent trips to Nepal have revealed 60 plant species used by local healers to treat 25 different diseases. Of these 60, only 28 had previously been reported to have medicinal activity, but in many cases for different diseases. Thirty-two species were not previously known to have any medicinal value at all.

The value of drugs and disease-resistant crops is easily measured, but other arguments for preserving biodiversity are less utilitarian. One is what may be called the 'aesthetic' argument, although the concept is broader than this. Put simply, biodiversity is beautiful, awe-inspiring or just plain interesting. One small part of this value of biodiversity can be estimated from the 500 million people worldwide who watched the 13-part BBC series *Life on Earth*, although of course most of the aesthetic benefits of biodiversity are experienced at first hand. One could also put a price on the new species introduced into gardens every year. Biodiversity may also have value merely by virtue of its existence, without being seen or experienced at all, even on film. Most of us are happier knowing that wild giant pandas or the Amazon rainforest exist, even if we do not expect to see either of them in the flesh.

Finally, there is the argument that transcends all others, which is the moral or religious view that we have no right to cause the extinction of other species, and that we should try to leave the Earth's biodiversity in more or less the same state as we found it. Many of us, even if we have never given conservation much thought, would be unwilling to bequeath a significantly depleted biosphere to our children and grandchildren.

As arguments for conserving biodiversity, all the considerations listed above have one big advantage, and one big disadvantage. Their advantage is that all of them suggest that we should at least try to conserve everything. Whether we are concerned with crop breeding or medicines, the value of species is to be found in their genes, and since at least some of the genes of every species are unique, who can say that any species will ultimately prove worthless? Equally, beauty is in the eye of the beholder, and any species, however ugly, venomous or just plain dull, may inspire awe, wonder or fascination in someone. And of course, the moral argument does not admit of shades of merit at all: if anything should be preserved, everything should be preserved. Mind you, there's an interesting argument that continues to this day about disease organisms. The smallpox virus is extinct in the wild and survives in only two heavily guarded medical laboratories, one in the USA and another in Russia. The World Health Organization has already changed its mind once about whether these remaining stocks should be destroyed. It's a sad commentary on the state of the world that the chief reason for keeping them is now the fear that clandestine stocks exist elsewhere and that these might be used in bioterrorism.

## Convincing the economists

The disadvantage of all the arguments advanced above is that economists do not find any of them convincing, and the world, by and large, is run by economists or by those who are advised by economists. Economists distrust anything without measurable value, which rules out the moral argument for a start - an absolute moral imperative cannot be valued in any way that makes sense. In any case, the moral argument for conservation may conflict (or at least appear to conflict) directly with another of equal or even greater moral power, for example that no one should go without food or a roof over their head. Potentially at least, aesthetic or 'existence' value can be measured. People may feel better because giant pandas exist in the wild, but just how much better? One way to find out (indeed the only way) is to ask: how much would you pay to preserve the giant panda? Or alternatively: if the giant panda were to become extinct in the wild, how little money would be required to compensate you for this? (One reason citizens of poor countries sometimes seem less concerned about biodiversity is that it takes relatively little to make them feel better off). In rich countries - which is where the question is usually asked - one difficulty is that the answers to such questions are notoriously suspect, if only because not everyone tells the truth. Pollsters routinely find that citizens of many countries are willing to pay more tax to provide better public services, but in the privacy of the polling booth they persist in voting for the party that promises the lowest taxes.

Whether to trust the answer is one problem, but economists can't even agree on the question. On the face of it, 'How much would you pay for a new nature reserve?' is just another way of asking 'How much compensation would you expect for the loss of a similar reserve?' But numerous studies show that the answer to the latter question ('willingness to accept compensation') is, on average, about seven times the answer to the former ('willingness to pay'). The reason seems to be that people value what they already have more highly than what they don't have. Worryingly, if we accept that people have a right to enjoy the natural world, then 'willingness to accept' is the right measure, yet 'willingness to pay' remains the usual approach. Nor is that an end to the problems of economic valuation. Research shows that even asking about monetary value can be a mistake; people are less likely to be altruistic and cooperative when asked to think purely in terms of money. Conserving a public good like the natural world is naturally a cooperative enterprise, but thinking about money brings out the selfish side in all of us.

Another vexed question concerns who has a right to an opinion on the value of nature, a problem illustrated by the story of the drug eflornithine. Eflornithine kills trypanosomes, the parasites responsible for African sleeping sickness. Even though it's by far the best treatment for this debilitating disease, its producer Aventis could not make a profit from selling the drug to poor Africans and abandoned making the drug for that purpose. Meanwhile, however, two other companies were profitably marketing eflornithine as a treatment to remove unwanted facial hair in women, and Aventis (together with Bristol Myers Squibb) agreed to resume production to treat sleeping sickness only after the charity Médecins Sans Frontières threatened to alert the public to the situation. Markets are very democratic instruments, but they do operate on the principle of one dollar, one vote, and are thus much better at meeting the unmet wants of the rich than the unmet needs of the poor - a problem that applies as much to the natural world as it does to life-saving drugs. In any case, even if the citizens of rich countries can agree about how much they value the biodiversity of poor countries, there is currently no practical mechanism for collecting and transferring the money from the former to the latter.

On the face of it, the potential marketable value of crop genes or new drugs provides the most economically legitimate rationale for conserving biodiversity, and at the same time both a rationale for and means of transferring resources from the rich world to the poor. After all, prospecting for oil has made Saudi Arabia rich and diamonds have made (a few) South Africans rich. Why shouldn't the chemical riches of the rainforest do the same for Belize or Ivory Coast? However, this has not happened on a large scale, and several arguments suggest it might never do so. One problem, paradoxically, is that every new wonder drug discovered in the rainforest reduces the value of what remains. If a chemical company already owns a goose that lays golden eggs, or maybe even a small flock of such geese, why waste time looking for more? A related problem is that once one area of forest is protected, containing many possible sources of new pharmaceuticals, the economic case for conserving any more forest is surprisingly small.

There are also scientific difficulties inherent in the way that plants and animals make chemicals. Living organisms do make chemicals that are pharmaceutically useful, but at only a very low frequency - much less than one in every 1,000 natural chemicals tested has any potential as a drug. Moreover, organisms are much cleverer than human chemists at making complex molecules, which means that even when a candidate compound has been identified, it may be impossible to synthesise (although modern molecular biology may mean that an easily cultured bacterium may be persuaded to make it). This can spell disaster if the origin of the chemical is a slow-growing plant. The cancer drug taxol was formerly extracted only from the bark of the Pacific yew, *Taxus brevifolia,* and the bark of four to six trees was required to treat a single patient. The Pacific yew might have been driven to extinction if recent research had not shown that related chemicals can be extracted from much commoner yew species in Asia and Europe. The rosy periwinkle *has* been all but exterminated in Madagascar. The 'ideal' plant-derived drug would be patented, thus providing income for its country of origin, but also capable of being artificially synthesised, thus relieving pressure on the natural source. Unfortunately such compounds are extremely rare. History teaches us that the most useful sources of drugs have been easily cultured microorganisms (e.g. the polyketide antibiotics penicillin and streptomycin). It's therefore perhaps not surprising that few drug companies spend more than a tiny fraction of their research and development budgets on bioprospecting. Finally, although the Rio Convention on Biological Diversity may improve matters in the future, relatively little of the value of compounds from tropical plants has so far found its way into the pockets of the people in the countries from which the plants originated.

## Valuing the biosphere

However, there are other ways of valuing the biosphere, and one in particular has generated much interest in the last ten years. This concerns what ecologists call 'ecosystem services' - put more simply, the planetary life-support system. All living organisms, but chiefly plants and microbes, work together to maintain the composition of the atmosphere, regulate the climate, provide clean water, control erosion, turn rocks into soil, fix atmospheric nitrogen, detoxify pollutants and generally make the earth habitable. What is all this ceaseless activity worth? In one sense this is a stupid question, since humans could not survive without these free services. They are literally both priceless and irreplaceable. Nevertheless, it is a question worth asking, if only in an attempt to make us appreciate just how much our continued existence depends on the continued healthy functioning of the planet. Also, although it is indeed pointless to ask what these services are worth in total, it *is* worth asking how relatively small changes in their quantity or quality might impact on human welfare.

The last major study that attempted to value the biosphere in this way, in 1997,[1] estimated that ecosystem services were worth somewhere between 16 and 54 trillion US dollars, with an average of about 33 trillion dollars at 1997 prices (a trillion is a thousand billion). That's such a large number that it's effectively meaningless, so let's focus down until we get some numbers we can grasp. Let's think just about the value of insects, a group whose conservation (compared with, say, birds or mammals) is not always given the importance it deserves. Furthermore, since we are primarily interested in natural biodiversity rather than farming let's stick to strictly wild insects: that means we ignore honey and silk and the vast industry devoted to raising insects for biological control. Sensibly, let's also stick to those services provided by insects where we have some hope of quantifying what they provide. Finally, we'll confine our attention to the USA: the data are better there.[2]

So, what are we talking about? Four services can, with some effort, be roughly quantified: dung burial, pest control, pollination and wildlife nutrition. Already you may be surprised by some of these. Dung burial? The fact that neither you nor anyone else has to even think about this, let alone worry about it, just goes to show what a fine job all those dung beetles are doing. Australians don't need telling this - before beetles that could deal with cattle dung were introduced, the Australian cattle industry was in danger of being engulfed by a rising tide of its own ordure. Dung beetles prevent fouling of forage, promote dung decomposition into useful plant fertiliser, and reduce the populations of parasites and pest flies. The value of this service amounts to $US 380 million annually at 2006 prices. That's impressive, but tiny compared with the other three services listed above. Control of pests by insects reduces crop losses every year by $4.49 billion, and crop pollination by wild insects is worth $3.07 billion. Even these, however, pale by comparison with the value of wildlife nutrition. US citizens spend over $60 billion every year on shooting, on fishing and on observing wildlife, and insects are a critical food source for much of this wildlife. The value of insects to birdwatching alone is $19.76 billion, and the total value is a whopping $49.96 billion. In other words, the value of these four services alone is estimated to be worth almost $60 billion to the US economy every year.

In reality, the figure is certainly higher. Insects do lots of other things (e.g. weed control, soil improvement, facilitation of decomposition of dead animals and plants) where it would be harder to work out the value of the benefits. And, of course, the figures apply only to the USA, and would be much larger if scaled up to the whole world. I should also point out, to keep the economists happy, that the 'values' described above refer to real financial transactions. If there were no dung beetles, forage fouling would reduce the value of rangeland and fewer cattle could be fed, slower dung decomposition would require more money spent on fertiliser, and more pests would mean more money spent on pesticides and medicines.

From this brief look at insects, it's not at all hard to see how the grand total for the value of ecosystem services reached $33 trillion in 1997 (which of course would be far more at today's prices). Much the largest part of this value concerns the functioning of the very fabric of the planet: fixing of atmospheric nitrogen (i.e. combining with other elements so it can be used by organisms), maintaining the cycles of other elements, waste treatment and pollution control, water cleansing, maintaining the balance between carbon dioxide and oxygen, and so on. To put the value of $33 trillion in perspective, it was at the time nearly twice global GNP. In other words, if we wanted to replace these services, we would have to almost triple global GNP, and having done that, we would be no better off: that impossible task would be necessary merely to end up where we are right now.

Other studies, some using quite different methods, have arrived at a similar figure for the total value of ecosystem services. Nevertheless, it's almost certainly too small by quite a wide margin, for several reasons. First, the authors of the study were only able to assemble the results of other smaller studies, and the values of some services and for some parts of the globe simply haven't been measured. Second, the authors were only able to extrapolate from the current marginal value of ecosystem services: in economic jargon, they were not able to put together a realistic demand curve for these services. In reality, of course, the value of such services would rise as the supply diminished - we can assume that the last breath of air and the last glass of clean water would be worth a lot more than they are now. Third, while the methods used to value ecosystem services were not uniform, they were mostly of the willingness-to-pay type. These values therefore reflect current public understanding of the value of the world's natural capital, which is certainly hopelessly inadequate. In a world where everyone was aware of just how much their daily existence depends on the continued delivery of ecosystem services, the value of those services would surely increase.

Finally, the valuation assumes that the world is an orderly, linear sort of place, with no sudden thresholds or discontinuities or, worse still, irreversibilities. In other words, there are no nasty surprises in store and, even as we continue to abuse the planet, we could (if we wished) put things right by simply retracing our steps. If experience tells us anything it is that the world is not like this, which means in practice that ecosystem services are worth more than we think they are: wouldn't you take better care of your present car if you knew there was a chance, however small and however hard to measure, that you would never be able to buy another one? For all these reasons, and indeed for several others, a mere $33 trillion is without doubt a major underestimate.

## Private profit, public loss

So, if ecosystem services are so valuable, why are we not falling over ourselves to conserve the ecosystems that provide them? The answer is that despite the oft-repeated phrase, there really *is* such a thing as a free lunch, and most of us receive one every day. In other words, the overwhelming majority of ecosystem services do not pass through any kind of market. No one has to pay for them, and they are freely dispensed to all, whether rich or poor, deserving or otherwise. As any economist will tell you, nothing is more likely to be grossly undervalued than goods and services that no one has to pay for. Ecosystem services are thus an excellent example of what Garrett Hardin called 'the tragedy of the commons' in 1968,[3] although the notion had already been around since at least 1833, and, like most good ideas, can be traced back to the ancient Greeks: Aristotle certainly knew about it. Hardin illustrated the idea by the metaphor of a field grazed in common by a number of shepherds, but crucially neither owned (or regulated) by anyone. The number of sheep on the common is exactly the number that it can sustainably support. Now, what is the rational behaviour of any one shepherd? If he adds another sheep to his flock, every sheep (and not only his) has just a little less to eat, but he gains more or less the whole advantage of the extra sheep, so it is clearly in his interests to increase his flock. Unfortunately the same applies to every shepherd, so more and more sheep are added until the common is destroyed. Note that no one has to behave badly or even stupidly for this to happen, for the tragedy is simply an inevitable consequence of the sum of a number of entirely rational individual decisions.

We can draw another conclusion from Hardin's parable. The world's rich nations long ago increased their flocks up to - and beyond - that which can be supported by the 'global commons', and are now keen to persuade poorer nations to show more restraint than they ever did.

So what would the world be like if ecosystem services were actually valued at their true worth? This is a difficult question to answer, not least because ecosystems converted to human use do not simply stop providing these services. In fact, substantial goods and services may still be obtained from forest *after* it has been converted to pasture or to a rubber plantation, which of course is why the conversion happened in the first place. What we need to know is how the services provided by agricultural or other human land uses measure up against the services provided by 'wild' nature before conversion by man. When scientists tried to assemble the evidence a few years ago, they found that surprisingly few people had asked this question. Only five relevant studies existed: two looked at tropical forest used for logging or converted to agriculture, one looked at conversion of mangroves to shrimp farming in Thailand, another at draining a Canadian marsh to provide agricultural land, while the last examined dynamite fishing of coral reefs in the Philippines.

The scientists who looked at these systems asked two questions. First, what was the original ecosystem worth, in terms of the ecosystem services defined above, plus the value of the sustainable products from that system (e.g. fruit, nuts, firewood, charcoal, sustainable hunting or fishing, tourism)? Second, what was the system worth after conversion, including ecosystem services and the private value of timber, crops, shrimps or whatever? It was thus possible to arrive at the total economic value of the five systems, both before and after conversion. In every case, the value after conversion was less than that before, usually by a wide margin. One system, tropical forest in Cameroon, even had a *negative* total value after conversion to oil palm and rubber plantations. Before we go any further, it is important to make it clear that nobody is suggesting that conversion of wild nature to human use has never been, or never will be, economically justifiable. Clearly society has benefited from conversion of forests and wetlands to agriculture: it is only because of such conversion that you and I have enough to eat. Nevertheless, it seems clear that further conversion of the *remaining* wild habitat does not make sense in terms of global sustainability.

Put simply, such further conversion does not make sense because the gains from the products of conversion are worth less than the losses of ecosystem services. In fact the economic costs of conversion are enormous. Over 1 per cent of the remaining area of wild habitat has been lost every year *since* the Rio Summit in 1992, losing Mankind plc $250 billion every year. So why does conversion continue? There are essentially three reasons, two of which have already been mentioned. First, a failure of understanding: people will not pay for something if they do not comprehend its true value. Second, conversion often *does* make narrow economic sense, because the gains of conversion are immediate, private and highly focused, while the losses are global, long-term and diffuse. Those who convert forests or mangroves obtain the benefits of the timber or shrimp harvest, but pay very little of the costs of climate change, erosion and flooding. Third, even the narrow economic benefits often depend on tax incentives and subsidies. In two of the five case studies mentioned, even the immediate private value of the conversion was negative without such 'perverse' subsidies. Depending on how you do the calculations, the global total of subsidies that are both economically and ecologically perverse is between $1 trillion and $2 trillion every year. If you think this sounds like a lot, consider that the European Union Common Agricultural Policy alone costs £30 billion every year, and that many subsidies are not nearly so obvious.

An interesting effect of many perverse subsidies is that as well as generally favouring the few at the expense of the many and the rich at the expense of the poor, they quite often have consequences that are the exact opposite of what was intended. For example, in the USA petrol has long been cheaper than bottled water (although that may at last be starting to change), which is only the most visible part of a subsidy to road transport that adds up to $1,700 per head of population every year (at 1996 prices). Subsidising oil consumption in this way discourages investment in new technologies and energy conservation and prolongs dependence on unreliable foreign supplies, which itself encourages the USA's unhealthy interest in the politics of the Middle East. Such subsidies, not surprisingly, arouse strong emotions: a 1999 book on the subject is subtitled *How Governments Use Your Money to Destroy the Earth and Pamper the Rich*.

However, we are getting ahead of ourselves. In the next chapter we will take a closer look at the impact of all this on biodiversity, and why current conservation efforts often don't seem to be working. We'll then consider perhaps the biggest question of all: What is the effect of biodiversity loss on the provision of the ecosystem services on which all our lives depend? If we have lost species (and we have), does this mean the planet now works less well than it used to, and if we continue to lose species (and we will), will it work even less well in the future? And what exactly do we mean by 'work' anyway? I must warn you that much of this is pretty depressing, although finally, in the last chapter, we'll see if we can find any cause for optimism about the future.

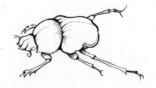

Dung beetle, *Scarabaeus sacer*

*Chapter 5*

# Threats to biodiversity

*We have inherited an incredibly beautiful and complex garden, but the trouble is that we have been appallingly bad gardeners. We have not bothered to acquaint ourselves with the simplest principles of gardening. By neglecting our garden, we are storing up for ourselves, in the not very distant future, a world catastrophe as bad as any atomic war, and we are doing it with all the bland complacency of an idiot child chopping up a Rembrandt with a pair of scissors. We go on, year after year, all over the world, creating dust bowls and erosion, cutting down forests and overgrazing our grasslands, polluting one of our most vital commodities - water - with industrial filth and all the time we are breeding with the ferocity of the Brown Rat, and wondering why there is not enough food to go round. We now stand so aloof from nature that we think we are God. This has always been a dangerous supposition.*

**Gerald Durrell,** *Two in the Bush* **(1966)**

## Clear and present danger

Almost everywhere, biodiversity is in retreat. This general fact may receive some publicity, but the biggest headlines are reserved for charismatic species facing obvious, clearly identifiable threats. For example, hippopotamus numbers have declined by at least 95 per cent in the Virunga National Park, a World Heritage Site in the Democratic Republic of Congo. Years of civil war have led to a breakdown in law and order, and armed factions are killing hippos not only for their meat but also for their canine teeth, worth a great deal of money in the illegal ivory trade. Blue whale numbers are currently about 3 per cent of their historical level, mainly as a result of hunting. Overfishing has reduced shark numbers in the north-west Atlantic by three-quarters in only 20 years. As coastal fisheries have declined, fishing has expanded into the open oceans, leading to a decline in tuna and swordfish of 80 per cent in just five years.

As the hippo example illustrates, biodiversity is often one of the first casualties of corruption, bad government and war, so it's a pity that biodiversity and war seem to go together. Between 1950 and 2000, more than 90 per cent of the world's major armed conflicts took place in countries containing the biodiversity hotspots discussed in Chapter 1, and over 80 per cent occurred within the actual hotspots themselves. Less than one-third of hotspots escaped serious conflict altogether during this period, and most suffered more than one war.[1]

Overfishing is just one example of how far people's ability to kill and harvest wild animals has outstripped their willingness to control their activities. The Maoris who wiped out New Zealand's moas, and the fishermen who removed countless millions of large fishes, sharks, sea turtles and manatees from the Caribbean in the seventeenth to nineteenth centuries (Caribbean turtle numbers have declined by over 99.9 per cent since 1500), at least had the excuse that they were unaware of the long-term consequences of their actions. Nowadays there is hardly a sustainable fishery anywhere in the world, and nearly all - on their present course - are destined to follow the Canadian Grand Banks cod fishery into oblivion.[2]

Sometimes a threat to biodiversity is direct but unintentional, and the threatened wildlife are just innocent bystanders. For example, diclofenac is a cheap and effective treatment for inflammation, pain and fever, widely used in human medicine globally, and introduced to the veterinary market on the Indian subcontinent during the early 1990s. Unfortunately, it persists for several days in carcasses of dead cattle, and is extremely toxic to birds. Its use has led to the catastrophic decline of several species of Indian vulture, whose main source of food is dead livestock. So severe was the decline that three species of vulture were facing extinction, and were saved (we hope) only at the last moment by the banning of diclofenac for cattle in 2006.

A remarkably similar story concerns the use of the broad-spectrum antiparasitic drug ivermectin in cattle, sheep, goats, pigs and horses. Residues persist in dung and have adverse effects on dung beetles and on other dung-feeding invertebrates, such as many flies. Again, once the problem is recognised, alternative treatments for parasites can be used. Both these examples are instructive, since in these cases not only are wild species threatened by veterinary treatment of domestic animals, but there is also a serious effect on vital ecosystem processes. We rely on the natural world to clear up dead animals and dung, and a decline in either dung beetles or vultures is bad news for human and animal health, nutrient cycling and agricultural productivity.

Cause and effect may not be so simple, and often all the links in the chain of ecological consequences can be traced only by painstaking detective work. The towering kelp forests of Western Alaska and the Aleutian Islands are one of the marvels of the marine world, and home to an array of unique animals. But in the last 20 years many of these forests have disappeared, taking the unique kelp ecosystem with them. Why has this happened? The beginning of the story seems to be (as usual) overfishing, leading to the collapse of fish stocks in the north Pacific and Bering Sea. This, in turn, has led to a dramatic decline in fish-eating seals and sea lions, which were themselves the main prey of killer whales. Logic suggests that the next link in this wretched chain should be a decline in killer whale numbers, but things are rarely that simple. Killer whales are intelligent, adaptable animals, who have responded to the decline in their normal prey by switching to sea otters. The result has been a collapse in otter numbers and an eruption of sea urchins, the otters' main food. Finally - are you still with me? - sea urchins graze on kelp, leading to a spectacular decline in kelp density.

Ironically, the change in killer whale behaviour has had such a sudden and catastrophic effect because sea otters are both smaller and less nutritious than seals (they have little or no blubber). So killer whales that have switched to otters have to eat large numbers: a killer whale that ate only sea otters would have to eat 1,825 a year. In fact the collapse of kelp forest over huge areas could have been caused by just half a dozen killer whales switching exclusively to feeding on sea otters. You have to feel sorry for sea otters: the fur trade almost led to their extinction by the early twentieth century; protection from hunting allowed a recovery of thriving populations throughout most of their historic range by the early 1970s - just in time to be hit by the latest disaster.

Oddly, sea urchins are also central to another well-documented example of ecological 'regime change'. Worldwide, corals are in competition with seaweeds, since both like to grow in warm, shallow, well-lit water. Corals generally prefer low-nutrient water, so *eutrophication* (an increase in chemical nutrients) of coastal waters through increased nutrient input from agriculture and soil erosion tends to result in domination by seaweeds rather than corals. However, even in the face of such pressure, grazing by a healthy herbivore community usually keeps the seaweed in check. In 1983, reefs around Jamaica shifted suddenly from domination by corals to domination by seaweeds. Centuries of overfishing of herbivorous fish had left the control of the potentially dominant seaweeds to a single species of sea urchin, *Diadema antillarum*. Sea urchin numbers had expanded to fill the gap left by the fish, until the urchin was hit by a pathogen and its population collapsed. As a result, Jamaica's reefs have shifted to a new, seaweed-dominated state which (unlike the original coral reefs) is both low in diversity and worthless as a fishery. 'Keystone species' is the name ecologists give to species, like the sea urchin in these examples, whose presence (or absence) can transform the very nature of an ecosystem, but we still have no better way of identifying them than taking them away and seeing what happens.

The complexities of these two marine examples have been revealed by years of dedicated research, but there must be many more cases where we are quite ignorant of why a species, or even a whole ecosystem, has suddenly declined or disappeared. Ecology continually teaches us that ecosystems rarely operate in a simple, linear fashion with causes leading to obvious effects. A related problem is that threats rarely operate alone, and a species decline may be driven by two or more threats working together. For example, in recent years few groups of animals have declined more severely than amphibians, a phenomenon probably attributable to the combined effect of climate change, pollution and disease. Humans may not only initiate the chain of disaster but also participate as victims further along the line. In Ghana, overfishing and the consequent decline of marine fish stocks has led people to turn to other sources of protein, in turn leading to an impact on mammal populations in nature reserves through illegal hunting.

# No way back

Our tale of two urchins also reveals another disturbing phenomenon, which is the tendency of many ecosystems to respond to external pressures not by changing smoothly and continuously, but by flipping quite rapidly between two or more stable states. The brown seaweeds that now dominate Jamaica's reefs not only prevent coral larvae from settling, they are also quite unpalatable to herbivores (which mainly eat young, tender algae), so the seaweed-dominated state seems to be here to stay. There may be no way back for Jamaica's corals.

Once you start to look, such dramatic ecosystem shifts are quite common. More than a decade after the complete collapse of the Canadian Grand Banks cod fishery, there is no sign of any recovery in cod stocks. The virtual elimination of cod seems to have pushed the ecosystem into a new, stable state dominated by crabs and shrimps. On land, there is evidence that woodland and grassy, open vegetation are often alternative stable states, especially in dry climates. In the absence of trees, it may often be simply too dry for tree seedlings to survive, especially in the presence of herbivores. In high-altitude, tropical 'cloud forests', water from clouds condenses in the canopy, keeping the whole system moist and often supporting extraordinarily diverse plant and animal communities. Cut down the trees, however, and water input stops and it can become too dry for the forest to recover. Many endangered or extinct amphibians are (or were) inhabitants of cloud forests. On a larger scale, desert and moist, vegetated conditions often seem to be alternative stable states. There is good evidence that the Sahel in north Africa flipped from a wet to a desert state about 5,000 years ago.

Finding that things break and then can't be fixed, and 'non-linearities' in general, is depressingly common in ecology. The only surprising thing is our continuing ability to be surprised by them until it's too late. We rarely see these disasters coming, but we are gradually getting better at it. Let me give you an example. Throughout the world, large areas of tropical forest have historically been managed by shifting cultivation ('slash and burn'), in which forest is cleared for agriculture, cropped for a few years, and then abandoned to allow the forest to recover before repeating the cycle. In both Asia and Africa, shifting cultivation is still the single largest cause of tropical deforestation. The expectation (or the hope, at any rate) was always that as long as the fallow periods were long enough, shifting cultivation was completely sustainable, i.e. you could keep on doing it forever. The trouble is, few scientists had ever investigated what happens after repeated cycles, and certainly none had looked at what happens beyond the first few cycles. The first really thorough, long-term study[3] (in Yucatán, Mexico) has recently revealed that forests subject to repeated cycles of clearance start to run out of phosphorus. This is bad news, because tropical soils are nearly always short of phosphorus to begin with. Worryingly, after only three cycles of clearance and fallow, there's too little available phosphorus left to regenerate a mature forest, although there is enough to grow a (lower-biomass) secondary forest.

What is going on here? It's tempting to speculate that the phosphorus is being lost when ash is blown away after burning, or by soil erosion or being washed out of the soil during the cropping interval, or removed in the harvested crop. One of the frustrating (and exciting) things about ecology is that such obvious answers are often wrong, and this is no exception. All the above processes do occur, but the main cause of the problem turns out to be reduced phosphorus deposition. Forests 'top up' their phosphorus reserves by trapping airborne dust, canopy condensation and fog. Young secondary forests don't do this very well, and fields of crops hardly do it at all, leading to a downward spiral in phosphorus content: forests with insufficient phosphorus can't grow as big, which means they trap less phosphorus, which means they have even less phosphorus . . . and so on. Calculations suggest that as few as six cycles might tip some forests into an alternative state in which there is simply too little phosphorus to generate a forest of any kind, leading to sparse, shrubby vegetation that might persist indefinitely. Whether this actually happens or not depends on how farmers respond to declining crop yields. Looking on the bright side (from a conservation perspective), they might lengthen the fallow periods or abandon farming altogether. Pessimistically, they might move on to new, virgin forest, or simply convert the worn-out secondary forest permanently into pasture, or both.

# Habitat loss

Overfishing, poaching, unintended effects of veterinary drugs: serious as these problems are, they are the easy ones to deal with. We can replace - and we largely have - diclofenac and ivermectin. Given enough money and political will, even if we couldn't solve the long-running strife in the Congo we could at least train and pay enough wardens to protect the hippos. We could for the first time actually listen to any of the authoritative forecasts of collapse of marine fisheries and set up some sufficiently large and effective marine reserves. Even if ending overfishing isn't the complete answer, it should at least give threatened marine animals a fighting chance of coping with pollution and climate change.

Unfortunately, the causes of much of the biodiversity crisis are both deeper and much harder to treat. For every species directly threatened by poaching or overharvesting, there are a hundred - or a thousand - that are just as endangered, despite not being the target of any specific threat. These species are simply being swept away by habitat loss. Tropical forests are by far the best-studied example, partly because of their extreme species richness (the total number of species in a given area) and partly because their extent can easily be monitored by satellite. Throughout the 1990s, tropical forests were lost at a rate of about 0.8 per cent per year, but in some areas it's much worse than this. The forests of Southeast Asia have been badly hit by a combination of burning, logging and replacement by oil palm plantations. The damage increasingly threatens former protected areas and has led to predictions that orang-utans may be extinct in the wild in a few decades. Loss of Southeast Asian mangroves - which are vital as fish nurseries and as storm protection - continued throughout the 1990s at a rate of about 1 per cent per year, although even that is an improvement on the rate of loss in the 1980s. On the Great Barrier Reef, coral cover has declined by about 30 per cent since 1965, while in the Caribbean coral cover has declined by about 10 per cent per year for the last 30 years. Global losses of wetlands, grasslands and freshwater are harder to monitor, but there's no reason to assume they are faring any better.

We know more about birds than we do about most other groups of animals or plants, and a survey in 2006[4] reveals that habitat loss or degradation is the main threat for around three-quarters of endangered birds, with losses to agriculture alone responsible for nearly half. Habitat loss, of course, is often merely the end product of a process that makes habitats increasingly less suitable for their original inhabitants. Sometimes these changes can be subtle, but devastating nonetheless.

In hindsight, it seems frankly incredible that there were serious proposals in the 1960s to build two massive dams within the Grand Canyon, flooding most of its length. Nevertheless, the Hoover Dam did flood the lower 20 per cent of the canyon and, more significantly, the Glen Canyon Dam lies some 25km upstream from the canyon's entrance. The visual effects on the canyon are obvious: seasonal floods no longer scour sand from the channel bed and deposit it on bars, eddies and beaches. Plants and animals that had evolved in the canyon for millions of years, and were uniquely adapted to the seasonal disturbance, now find their home a much-altered and far less congenial place. But there's another more subtle effect. The Colorado River in the canyon now derives from the bottom of a deep, cold lake. So the water, which used to vary from 2°C in the winter to 29°C in the summer, is now a chilly 7°C year-round. Native animals, adapted to the warm summer water, are declining or gone: the Colorado pikeminnow, the bonytail and the razorback sucker are gone from the canyon, while humpback chub numbers are down by 80 per cent. No one has seen a southwest river otter for over 30 years and it is almost certainly extinct. And if you find it hard to get worked up about a minnow, the Colorado pikeminnow is the largest minnow in America and one of the largest in the world, at almost two metres long.

We know a lot about the environmental effects of the Glen Canyon Dam, because it happens to be in one of the world's great tourist attractions, which in turn is in the world's richest country, with plenty of money to spend on endangered wildlife. There are plans to save all four of the fish mentioned above, and to reintroduce other otter subspecies to the Colorado (but it's worth noting that even here there have probably been extinctions of invertebrates that have either gone unnoticed or are unlikely to provoke any major conservation effort). Globally, half the world's rivers now have at least one large dam, and there are at least 45,000 large dams worldwide, flooding more than 450,000km². One of the most insidious effects of dams is on the distinctive local seasonal pattern of erosion and deposition, which varies from region to region, and even from river to river, and to which unique local animals and plants are precisely adapted. Dams destroy all this local environmental diversity: all dammed rivers are rather similar, favouring the spread of a few cosmopolitan, adaptable species at the expense of the unique local biota. China's Three Gorges Dam certainly hasn't helped the Yangtze river dolphin, of which no trace was found in a major search in 2007, although pollution was probably a bigger cause of its demise. The dolphin, which had been around for 20 million years, is the first cetacean to have become extinct as a direct consequence of human activity.

People are easier to count than dolphins, and we know between 40 and 80 million people worldwide have been displaced by dams.

# Habitat fragmentation

Complete habitat loss - the conversion of tropical forest to oil palm plantation, soy bean cultivation or cattle pasture, say - may be the worst that can happen, but even a habitat that persists (and may even be protected) may fail to continue to provide suitable conditions for many of its original inhabitants. An interesting case study is Wolong Nature Reserve in Sichuan, south-western China. Wolong is important by any standards: it's the largest giant panda reserve (200,000ha), with 10 per cent of the wild panda population. It's also a flagship of the Chinese government's conservation programme, and has received considerable support, both financial and technical, from organisations such as the World Wide Fund for Nature (WWF).

The reserve was designated in 1975, and high-quality panda habitat has declined dramatically since then. Essentially this is due to continuing deforestation: forest fragments have continued to shrink and disappear, and formerly extensive tracts of forest have become divided into smaller patches. The reasons are not hard to find. The human population in the reserve nearly doubled between 1975 and 1995, and the number of households more than doubled. Economic activities within the reserve include farming, fuelwood and timber harvesting, Chinese herbal medicine collection and tourism. A boom in tourism has stimulated the extraction of fuelwood to make marketable products, leading to exploitation of more remote forests at higher elevations (often prime panda habitat). The effects on the panda population have been predictable: there were 145 pandas in 1974, 72 in 1986, and although up-to-date figures are not available, probably fewer today. To a large extent, the fate of Wolong is typical of the failure of many reserves successfully to resist the increasing pressures of a rising human population.

Sometimes, even though much intact habitat remains, the individual patches are just too small. Although ecologists had long suspected this to be the case, it was hard to prove until the Biological Dynamics of Forest Fragments Project was set up in the early 1980s by far-sighted ecologist Thomas E. Lovejoy. The project is an experimental 1,000km$^2$ landscape 80km north of Manaus in central Amazonia, in which forest fragments ranging from 1ha to 100ha (all at least 100m apart) were created by clearing and burning the surrounding vegetation to create cattle pastures. The beauty of this experiment is that changes in each fragment have been monitored since they were created, and also fragments can be compared with nearby continuous forest.

We'll look at plants in a minute, but first, what happened to birds? Loss of species following fragmentation of the forest is extremely rapid, and the smaller the fragment, the faster the loss. Indeed, loss of some species is so fast that it's only by sampling from the very moment the fragments are created that one can be sure the species were ever there at all. Fragments of 1ha lose half their bird species in less than ten years, 100ha fragments take around one or two decades to do the same, and larger fragments take a few decades to perhaps a century. Roughly, to increase tenfold the time it takes a fragment to lose half its birds, its area has to be increased a thousand-fold. For conservationists, this is crucial information. It's a fundamental and long-known law of ecology that, all else being equal, larger patches of habitat contain more species than small ones - and much head-scratching has been devoted to the question of how large a forest fragment needs to be to provide useful habitat for forest animals. Missing from much of this debate, however, has been the critical element of time. That is, how big must a fragment be to retain most of its species indefinitely, or at least until they can be 'rescued' by regrowth of surrounding forest? Given that such regrowth takes 100 years to produce anything like the original forest, patches of around 1,000ha (10km$^2$) are the absolute minimum required. If - as is often the case - there is no hope of forest regrowth coming to the rescue of the stranded species, fragments need to be much larger. Even 10,000 ha (100km$^2$) fragments would lose many bird species (starting, inevitably, with those at greatest risk) if isolated for a century.

The fact that fragmentation was so disastrous for wide-ranging, mobile animals such as birds came as no surprise to most ecologists. On the other hand, trees were expected to be affected less, and more slowly. Up to a point, this proved to be true: species richness of trees did not change, even in small fragments. But there were many rapid and highly significant changes in species composition. Big, slow-growing, shade-tolerant trees declined dramatically, especially near forest edges, and were replaced by smaller, fast-growing opportunist species. The main cause seems to be increased exposure of forest edges to fire, drought and storms. But there were changes even in the interior of larger fragments, suggesting that disruption of populations of pollinators and dispersers was also involved. Crucially, trees favoured by fragmentation had much lower wood density than disadvantaged species, so fragments store much less carbon than similar areas of intact forest. It seems clear that to survive in the long term, forest trees need just as large intact areas as do forest birds.

The Forest Fragments Project is a deliberate experiment, but every now and then a similar research opportunity turns up by accident. One such project is the creation of Lago Guri, a 4,300km$^2$ lake created by a hydroelectric dam in the Caroni valley in Ecuador. The rising waters have isolated a number of islands, some very small, and all covered by tropical forest. Because some of these new islands are so small (just a few hectares), by chance they lack most larger vertebrates, especially predators that are normally found at quite low densities. Similar areas of mainland, on the other hand, have a full complement of cats, mustelids (ferrets, badgers, otters, etc.), large birds of prey and large snakes. But on the islands, released from their predators, herbivore populations have exploded. Rodents are 35 times, iguanas 10 times and howler monkeys 30 times more abundant than on the mainland. There is a predictable knock-on effect on the vegetation: saplings of canopy trees are only 20 per cent as abundant as on the mainland. We aren't yet very far down the road created by the formation of the lake, and we can expect events to take a few more decades to work themselves out. But the final state seems likely to be the elimination of the original diverse forest and its replacement by an odd collection of spiny, toxic and otherwise herbivore-resistant plants. Along the way, much of the original animal diversity will probably disappear too.

A glimpse of what that might look like can be gained from Quebec's Anticosti Island, a 7,943km$^2$ island in the Gulf of St Lawrence.[5] It's not tropical, and we're looking at the introduction of a herbivore rather than elimination of predators, but the principle is the same. Anticosti was originally deer-free, but 220 white-tail deer were introduced in 1896. Today there are around 100,000. Not surprisingly, the island's vegetation has been transformed. Since all the shrubs the deer like to eat have gone completely, the deer have moved on to balsam fir, normally eaten only as a last resort. Every fir within reach has been devoured, driving the forest towards a monoculture of white spruce, which the deer won't eat at all. Yet even today the deer persist, so what on Earth are they eating? The answer is balsam fir twigs and lichen that fall out of the canopy during storms, but even this won't last for ever. As the balsam firs die out, there will finally be nothing to eat and many of the deer will starve, yet only after causing damage that will be hard to reverse. One casualty of the deer introduction has been black bears, once abundant but now extinct since the berrying shrubs they depended on have been wiped out. Since deer hunting is the basis of the island's economy, steps are now being taken to fence off large areas to allow balsam fir to regenerate, but the bears, shrubs and other browsing-intolerant plants won't return. Anticosti is now essentially a large deer farm.

Lago Guri and Anticosti illustrate an important principle, which is that plants are not normally wiped out by herbivores for one very good reason: before that can happen, the herbivores are controlled by their predators. An ecosystem without predators is an unnatural and usually an unhealthy ecosystem. You may think Lago Guri and Anticosti are extreme examples, but large predators are often among the first species to be lost when habitats are lost or fragmented. Even ecologists who thought they knew what to expect have been astonished by the resurgence of plant and animal diversity in Yellowstone National Park since the reintroduction of wolves in 1995. Trees such as aspen, previously an important component of the park's vegetation, were being killed or damaged by an overabundant elk population. Now that elk are being controlled by wolves, aspen and many other herbivore-sensitive plants are recovering in the park. There is evidence that wolves could do the same in many other ecosystems. In Scotland, red deer support a trophy-hunting industry but their overall economic impact is negative, since overgrazing by deer hampers forestry, competes with livestock and reduces densities of many birds. Recent research[6] suggests that wolves, extinct in Scotland since 1769, could have an overall positive economic impact if reintroduced, although some farmers are yet to be convinced.

Nor do the predators need to be particularly large or charismatic. A 2006 study[7] in the Italian Alps compared sites with territories of either goshawks or several owls with similar sites lacking such territories. Sites with resident raptors had more plant species, more bird species and more butterfly species, as well as higher densities of both birds and butterflies. It's hard, without further research, to uncover all the direct and indirect causes of this relationship, but now you know why lions always look so smug: they *know* how indispensable they are.

The Biological Dynamics of Forest Fragments Project has taken 20 years and involved an immense amount of work: the capture, identification and ringing of over 21,600 birds, and the monitoring of the fate of nearly 32,000 trees of 1,162 different species. Yet, remarkable as this is, trees and birds are two of the few groups for which it would have been possible at all. Most invertebrates, particularly in the tropics, are simply too diverse and too poorly known for their response to fragmentation (or to anything else) to be determined. Let's look at one example from Cameroon,[8] where a team of ecologists attempted to determine the effect of forest clearance on birds, butterflies, beetles, ants, termites and soil nematodes. The method was to examine a gradient from forest that had never been cleared (primary forest) through forest developed after cultivation (secondary forest) and *Terminalia* plantation to complete clearance.

Despite expending about five person-years on sampling, sorting and cataloguing, the team was forced to the conclusion that it had sampled adequately only the birds and butterflies. In all the other groups, not only were the inventories certainly incomplete, but around half of the captured species could not be assigned to any previously described species. Bearing in mind that no attempt was made to sample many extremely species-rich groups, the total number of animal species present is probably close to 100 times the approximately 2,000 species sampled. In other words, assessing the effects of partial or complete forest clearance on the full range of species present *at a single site*, in a reasonable time, could easily absorb 20 per cent of the entire global workforce of professional taxonomists. Is it any surprise that the true extent of tropical biodiversity, and of its rate of loss, remains almost completely unknown?

## Too much of a good thing

Unfortunately, even large areas of intact habitat are not necessarily proof against some insidious forms of human damage. Ever since the beginning of the industrial revolution, we have been pumping ever-increasing quantities of pollutants into the atmosphere. Nitrogen, required in large quantities by all life, makes up four-fifths of the atmosphere, but must be 'fixed' (combined with carbon, hydrogen or oxygen) before it can be used by most organisms. No higher plant can fix nitrogen, but many have teamed up with symbiotic bacteria that do the job for them. On land, the pre-eminent nitrogen fixers are the legumes and their *Rhizobium* bacterial partners, although several other unrelated plants have different nitrogen-fixing partners, including alder and the actinomycete *Frankia*. Before human intervention, around 100 million metric tons of nitrogen were fixed biologically every year on land, and perhaps the same again in the oceans. By artificially fixing nitrogen to make fertilisers, burning nitrogen-containing fossil fuels and growing large areas of leguminous crops, mankind has at least doubled the amount of fixed nitrogen in circulation. Much of this nitrogen pollution is gaseous, in the form of oxides of nitrogen or ammonia, and therefore its effects are extremely widespread. It would be surprising if this enormous extra nitrogen input had not had some equally large effects on the natural world, and it has.

In Britain, it's long been the custom for individual counties to publish *floras*, or accounts of the distribution of their plants, generally with maps in the more recent examples. Writing county floras has long been considered a suitable occupation for retired clergymen. Because such floras are occasionally revised or completely rewritten, a comparison of floras written at different times can reveal patterns of change over time. Such a study of the UK's county floras reveals that throughout the twentieth century, each county lost wildflower species at the rate of around one per year, and the loss continues today. Mostly these were not very rare species, so they're still to be found somewhere, but nevertheless the floral diversity of the British countryside is slowly ebbing away. Some of the loss is certainly due to the destruction of habitat: drainage of wetlands; replacement of old grasslands by arable agriculture, of old woodlands by conifer plantations, and of plant habitats of all kinds by tarmac and concrete. But that's far from the whole story. The pattern of loss is not random, with short, slow-growing plants of open, low-nutrient habitats far more likely to have become locally extinct. This points strongly to a blanket effect of increased fertility, or eutrophication, with atmospheric nitrogen deposition one of the most likely causes (although there are other sources of surplus nutrients, including intensive farming). Another effect of nitrogen deposition (sometimes also called 'acid rain') is acidification of soils. Both effects - eutrophication and acidification - combine to reduce local plant diversity, as we saw in Chapter 2.

A study in 2004[9] of one very widespread kind of British grassland, particularly common in upland sheep pasture, was designed specifically to reveal the effects of nitrogen pollution by sampling the full spectrum of rates of deposition, from less than 10kg per hectare per year (on some coasts and in northern Scotland) to more than 30kg per hectare per year (in the Pennines and south Wales). Astonishingly, a typical patch of grassland receiving the highest rate of nitrogen had, on average, less than half the number of plant species of patches receiving the lowest rate. For every extra 2.5kg of nitrogen per hectare per year, a species is lost (and research in 2008[10] showed that American prairies respond to nitrogen pollution in exactly the same way). Most of us are familiar with the disastrous impacts of intensive agriculture on biodiversity, but mostly these impacts involve the original habitat simply being swept away, e.g. the felling of ancient woodland or the ploughing and fertilising of old grasslands. The particularly depressing aspect of the damage caused by nitrogen pollution is that has taken place in habitats (e.g. upland sheep pasture) that appear, on the face of it, to have remained largely undamaged.

The sobering fact is that anyone born in Britain in the second half of the twentieth century has grown up with, and come to accept, a floristic landscape that would have seemed oddly, inexplicably and depressingly dull to anyone born 100 years ago. Your grandparents are right: things really aren't what they used to be. One consequence of this dramatic (but on a human timescale, almost imperceptibly slow) loss of diversity is a kind of conservational 'mission creep': we now leap to the defence of patches of vegetation that wouldn't have merited a second glance 100 years ago. It's worth mentioning, by the way, that much the same process has occurred in our coastal waters: if you want to know what marine biodiversity *should* be like, ask any sea fisherman over the age of 50.

An odd consequence of this saturation with nitrogen is to make European vegetation extremely vulnerable to the addition of extra phosphorus. Nitrogen and phosphorus are the big two: the elements that all life needs in large quantities, and throughout history soil fertility has often been limited by a shortage of one or the other (or both). Massive nitrogen deposition removes one of these brakes, and as a result the fertility of many modern soils, drenched in nitrogen, can be transformed by even quite modest amounts of phosphorus. Recently, Czech researchers returned to a short-lived experiment in which fertilisers were briefly applied to infertile grassland in the mid-1960s. Now, over 40 years later, the vegetation of plots that received even quite small amounts of phosphorus is still completely different from plots that received no fertiliser, or nitrogen alone. This extreme persistence of phosphorus in soil is a serious barrier to the conservation or restoration of any kind of vegetation that needs low-fertility soils (which, as we've seen, includes most kinds of diverse and interesting vegetation).

From these examples it's easy to get the impression that nitrogen deposition is a problem confined to the developed world. It certainly began there, and that's where its effects have been most studied, but the rest of the world is rapidly catching up. Atmospheric nitrogen deposition is set to increase dramatically worldwide in the coming decades as China, India and others burn more coal and drive more cars. In a particularly fine example of Murphy's Law, nitrogen pollution in the world's biodiversity 'hotspots' will be around 50 per cent greater than the global average. In one such hotspot, the Western Ghats of India and Sri Lanka, nitrogen deposition could be 33kg per hectare per year by 2050, and in several others it will exceed 20kg. Note that 33kg is twice the average current rate of deposition that has already wreaked such havoc in the UK and central Europe.

Because there has been so little work on nitrogen deposition outside Europe and north America, we literally have no idea what the biological effects will be. In particular we have no idea what the effects will be in the tropics, but we can expect severe consequences both from nitrogen's fertilising effect and from its acidifying effect. Many tropical soils are already acidic and are therefore particularly poorly equipped to resist further acidification - a process already strongly implicated as one factor in the decline of many amphibians. We can also expect much terrestrial nitrogen deposition to be exported to lakes and rivers, where its effects will probably be just as disastrous as they have been already in northern Europe.

# Effects on humans

Current and likely future rates of biodiversity loss are bad enough purely on their own terms, but there are inextricable links between biodiversity and human well-being. Everywhere you care to look, damage to species and ecosystems adds up to shooting ourselves in the foot. I've already pointed out that converting mangroves to shrimp fisheries doesn't make economic sense, but careful study of the Indian Ocean tsunami of December 2004 reveals that there was less damage to property, and fewer deaths, in places where mangroves remained intact and healthy. The tsunami caused by hurricane Nargis, which caused such loss of life in Burma in May 2008, was also more damaging than it should have been, for exactly the same reason: indiscriminate felling of coastal mangroves, this time largely to grow rice. Smoke from the (largely deliberate) burning of 50,000km$^2$ of Indonesian forest in

1997 affected 70 million people, 12 million of whom required health care as a result. The global economic cost for that single year has been estimated at $4.5 billion. Increases in numbers of rats and feral dogs, leading to more human cases of rabies and plague across the Indian subcontinent, have been blamed on the catastrophic decline in Indian vultures due to poisoning by diclofenac. Deforestation nearly always causes human hardship: in one example, slash-and-burn by upland farmers in Madagascar led to increased river siltation and reduced water flows to 2.5 million downstream rice farmers. The economic impact was so severe that the Malagasy government decided in 2003 to triple the size of Madagascar's network of protected forests. Sometimes the economic loss can have global implications. Operation of locks means that each ship that passes through the Panamá Canal needs 52 million gallons of fresh water. This water comes from two lakes created by damming the Chagres River, but sedimentation caused by deforestation has reduced water availability. As a result, new reservoirs will be needed, flooding a further 50,000ha and displacing 8,000 inhabitants.

As human farmers, hunters and loggers press further into increasingly remote areas, they come into close contact with other species for the first time, and bring other species together, with increased risk of emergence and spread of novel diseases. For example, SARS, which caused such global panic in 2002 and 2003, seems to have spread to humans from Himalayan palm civets, but the civets were infected by finding themselves alongside bats – the original source of the SARS virus – in Chinese wildlife markets. It's estimated that 75 per cent of all human diseases that have emerged over the last few decades have come from animals, including SARS, AIDS, avian flu, Ebola, monkey pox and West Nile Virus.

So-called emergent diseases may generate the most newspaper headlines, but nothing compares with malaria, which kills over a million people every year, most of them children in Africa. Unfortunately, the anopheline mosquitoes that spread malaria thrive on deforestation. In Trinidad, a malaria epidemic followed the spread of the local mosquito carrier *Anopheles bellator* after forest clearance. In the Brazilian Amazon, the efficient malaria carrier *A. darlingi* was found in almost all sites altered by dams and roads, but was completely absent from undisturbed forest. In north-eastern Peru, determined efforts to eradicate malaria had virtually eliminated *A. darlingi* and reduced human cases to a few hundred a year. But with the completion of a road through the forest from Iquitos to Nauta, and the associated forest clearance and human settlement, *A. darlingi* reappeared in large numbers, and there were 120,000 malaria cases in 1997.[11]

Deforestation can also benefit the malaria parasite itself. The most dangerous form (there are four distinct species) cannot develop at all below 18°C, and development within the mosquito is dramatically accelerated by every degree above 20°C, so mosquitoes become infectious faster in open, sunny sites. The mosquitoes benefit too: one study in Kenya found that the time needed for *Anopheles gambiae* (the main African carrier) to lay eggs was 50 per cent less in open, treeless areas compared with nearby forest. Throughout Africa, a lethal cocktail of climate change, human population growth and deforestation is leading to the expansion of malaria into highland areas formerly free of the disease.

# Living on borrowed time

As we've seen, it's a fundamental law of ecology that small isolated patches of habitat contain fewer species than larger ones. This is why, eventually, if we turn most of the world's wild places into cities and intensive agriculture, we will also lose most of the species that used to live there. However, we've also seen that this doesn't happen overnight, and although the final outcome is not in doubt, we can continue to have higher diversity than we deserve for many years. Bearing this in mind, we can take a fresh look at the 25 global biodiversity hotspots that we first met in Chapter 1. Taken together, these hotspots, despite occupying just 1.4 per cent of the Earth's land area (formerly nearly 12 per cent), are home to the *entire* ranges of 44 per cent of the world's plants and 35 per cent of all terrestrial vertebrates. Never have so many eggs been left in so few (or so fragile) baskets.

The figures to focus on are the former area of 12 per cent and the current area of 1.4 per cent. Given these huge losses, we would expect that the hotspots would have already seen very many extinctions of endemic species (i.e. species that live nowhere else - we wouldn't expect habitat loss in hotspots to tell us anything about extinction of species that also occur *outside* hotspots). For a few groups that are reasonably well recorded (plants, mammals and birds) we can therefore predict how many hotspot endemics should have become extinct. If we do the calculations, there is some (qualified) good news, and some very bad news. The good news is that the hotspots have lost far, far fewer species than habitat loss would predict. The bad news is that habitat loss does quite a good job of predicting the total number of extinct *plus threatened* endemic species. The situation is patchy: the Caribbean, for example, has already lost over half its endemic mammals, and all the survivors are threatened. At the opposite extreme, only a quarter of the mammals endemic to the south-central China hotspot are threatened, and none (so far) are extinct. The difference here is largely that Caribbean mammals generally started out with smaller geographical ranges than Chinese ones, so were closer to extinction to begin with. The key point, however, is that we can confidently expect *all* threatened hotspot endemics to become extinct *because of habitat loss that has already taken place.*

We don't know exactly how long these extinctions will take, but the iron law that relates numbers of species to land area says that they surely will happen. I know this is quite a hard idea to swallow, but when extinction has been given enough time to occur, the results show that the predictions are right. Take, for example, the endemic birds of the forests of eastern North America. European colonists and their descendants began to clear these forests from 1620, and the low point was reached in about 1870, when about half the forests remained (since then, there has been a mild recovery owing to the abandonment of agricultural land). The calculations say that a 50-per-cent forest loss should mean extinction of about 15 per cent of the region's 30 endemic birds (the relationship is not a simple one: if you lose half the forest, you don't lose half the forest birds). That's about 4.5 extinctions. In fact three species have become extinct, and two others are critically endangered. One, the ivory-billed woodpecker, was last seen in the 1940s and was presumed extinct until apparently sighted in 2004. There is even a brief video of the bird, but not everyone is convinced it's not the commoner pileated woodpecker, and subsequent extensive searches have failed to find any conclusive evidence. The search continues, but the key point is that the number of extinct or almost-extinct species is in close agreement with theory.

Predicted extinctions in hotspots would be bad enough if habitat loss had now been halted, but it hasn't. For tropical forest hotspots, we can estimate the consequences of projecting current rates of deforestation into the future. Again, the Caribbean looks to be in particular trouble, largely because of high rates of deforestation in Haiti and Jamaica: here we can expect just five years' more deforestation to cause the eventual extinction of another 25 endemic vertebrates. The tropical Andes, Central America and the Philippines will also see many more extinctions at current rates of deforestation.

Clearly, to prevent an exceptionally large spasm of extinction, habitat loss in the hotspots (but not only there) must not only be halted but actively reversed. Given that urgent need, how are we doing? How good a job is being done by the current global network of protected areas? The complete analysis needed to answer that question hasn't been done on a global scale, but it has for the continental western hemisphere. The general conclusion is that most reserves are both too small and in the wrong place. Despite the concentration of biodiversity in the tropics, 35 per cent of the protected area is in Alaska. Of the 1,413 reserves, well over half are less than 10km$^2$, and the median area (i.e. the area at which half are larger and half are smaller) is just 4.86km$^2$. So little Atlantic Coast Forest remains in Brazil, and the remaining fragments are so small, that nowhere - even in the centre of the largest reserve - is it possible to get more than 12km from the forest edge. Recall the Amazonian fragmentation study that concluded that 10km$^2$ is the absolute minimum reserve size, even on a temporary basis, and that a reserve intended to provide a permanent refuge should be much larger than that. Not surprisingly, around three-quarters of threatened species are inadequately represented within reserves.

Not only are most reserves too small, they are also too far apart, and the land use separating them is too hostile to allow wildlife to disperse between them. Even quite small fragments of woodland can provide for the needs of a good deal of forest wildlife, if they are not too far apart and connected by the right kind of countryside. If the whole network is connected to a surviving tract of continuous forest, so much the better. A good example from Costa Rica is described in Chapter 7.

## The real size of nature reserves?

The analysis reported in the preceding paragraphs uses information from the United Nations official database of protected areas, which may well be far from complete. This may be good news: there may be more reserves than those in the official list, and of course there are landscapes where sympathetic management provides opportunities for wildlife, without being 'official' nature reserves at all. On the other hand, it does assume that all the areas listed are actually protected - for example, the 2003 list shows Wolong Nature Reserve as 200,000ha of UNESCO-MAB [Man And Biosphere] Biosphere Reserve, i.e. in the top bracket of international reserves specifically designed to "promote solutions to reconcile the conservation of biodiversity with its sustainable use". See page 73 for more about Wolong's failure to do this.

How unusual is Wolong? Or to put it another way, how protected are 'protected areas'? The first thing to note is that in many cases, it's simply impossible to say. In the huge tropical forests of both the Amazon and Congo there are enormous protected areas. Most of these are remote, and in practice they would be pristine wildlife habitat (i.e. they would be *de facto* reserves) even if they had no official protection. The value of their designation as reserves, in other words the ability of that designation to protect them in the face of an actual threat, is simply untested. You would have to say, however, that the signs are not encouraging. In the Amazon, for example, major new infrastructure projects are planned, and history teaches us that once roads are built, deforestation inevitably follows.

When a reserve is actually threatened, its protection is only as good as enforcement can make it. Or, as the late marine biologist Geoff Kirkwood put it (tongue only slightly in cheek), a reserve is only as big as the probability that a poacher will be caught and punished. That is, if the chance of detection is only 50 per cent, the reserve is only half as big as you think it is. To take an extreme case, in the mid-1980s the probability of a poacher being caught in the Serengeti National Park declined to around 1 per cent, effectively reducing the reserve from 25,000km$^2$ to 250km$^2$; the rest was just a free-range hunting reserve for poachers.

A study of poaching reveals a number of interesting (and uncomfortable) truths for conservationists. One is that increasingly harsh sentences are no deterrent, owing to what economists call 'discounting': essentially the value people put on future income relative to present income. In much of Africa, where people are poor and life is short, future income is heavily discounted and the possibility of a long prison sentence is no deterrent. On the other hand, a high probability of detection *is* a deterrent, because it results in *immediate* loss of income. This is only an extreme example of a problem that afflicts law and order worldwide. Calls for longer prison sentences come from the comfortable middle classes, who expect to live long, prosperous lives and thus do not discount the future at all, but are applied to criminals who are often poor, addicted to drugs or mentally ill and hardly consider the future at all. Thus long sentences do not deter, but those who want them are the least well-equipped to understand why.

Poaching also teaches us another lesson: be careful before applying standard economic theory to conservation. Economists tend to assume that human ingenuity and market forces always lead to the development of cheaper and more widely available consumer goods - think pocket calculators in the 1970s or mobile phones more recently. In other words, such goods start out scarce and highly prized, but then become cheap and commonplace. Unfortunately, many wildlife resources, for example ivory, do exactly the opposite - they start out common and unappreciated and become overexploited and expensive. Thus poaching in the Serengeti led to the local extinction of rhinos and severe depletion of the elephant population, while the price of rhino horn and ivory went through the roof. And if it seems difficult to draw any general lessons from the price of ivory, consider ivory as a metaphor for 'wild' land, the species that live there and the services they provide.

Maybe you think poaching is a remote, tropical problem? Well, perhaps it is, but wildlife in the rich world suffers from 'poaching' too. In Banff and Jasper National Parks in Alberta, Canada, large numbers of bears, moose, deer, coyotes and smaller animals are killed every year in collisions with motor vehicles and trains. From the point of view of the animals killed, this is effectively poaching, and the problem is exacerbated by the widespread failure to obey speed limits - which are rarely enforced. A partial solution is to build overpasses or underpasses for game, although ironically the main justification for the necessary expenditure is the reduction of human casualties in collisions. Large carnivores are also killed in direct conflicts with armed hikers, dog walkers and mountain bikers, who are encouraged by the extensive road network. The lesson, as in the Amazon, is that really serious wildlife conservation needs at least some large, road-free reserves.

As for the Serengeti, the chance of poachers being caught is now about 10 per cent - not perfect, but a big improvement. Better still, alternative sources of food and employment around the park have reduced the incentive for people to turn to poaching in the first place.

# Biofuels

Historically, the main driver of habitat loss has been the growth of crops for food, but there is a new kid on the block: biofuels. As oil and gas reserves dwindle and demand from developing countries grows, the world's attention has suddenly focused on the possibility of growing fuel and particularly on ethanol, which can easily be used to fuel vehicles. Brazil, the world's largest sugar producer, now devotes half its production to ethanol manufacture. This is one reason, but not the main one (which is the continuing - and extremely unhealthy - high demand for sugar in food), that sugar prices are at a 25-year high. Sugar is not a major ethanol crop in other countries, but we may expect that to change as production catches up with demand.

Other countries prefer to turn grain into ethanol. Driven by government subsidies and the runaway price of oil, ethanol production from maize is booming in the USA. As I write this, one-third of the US maize crop is already being turned into ethanol, and by the time you read this, that share will be even higher. However you look at it, the economics make no sense. The grain needed to make enough ethanol to fill the tank of a typical gas-guzzling SUV could feed one person for a year.

Even if the entire US grain harvest were turned into fuel for cars, it would still satisfy only one-sixth of US demand. And if turning the whole US grain harvest into fuel sounds ridiculous, bear in mind that the US Senate in 2007 passed an energy bill setting a target for ethanol production by 2020 that would require *more* than the entire current US maize crop. No thought seems to have been given to where all this ethanol will come from. And things aren't much different in Europe. A new British refinery will turn 1m tonnes of wheat into 400m litres of ethanol every year. That's about 7 per cent of the UK wheat crop, and about half of 1 per cent of the UK's motor fuel. To achieve the EU target of 5 per cent biofuel content in petrol, ten such refineries will be needed, and if they are fuelled by wheat they will consume 70 per cent of the UK crop. You didn't need to be much of an economist to imagine the impact of all this on world grain prices, although many people do seem to have been surprised.

This sudden dash for biofuels has two unwelcome consequences. First, it puts the two billion people who already didn't have enough to eat in direct competition with the world's motorists. Second, the extra demand for crops for biofuels will inevitably increase conversion of forests into croplands. Indonesia's plan to triple the area under oil palm by 2020, largely driven by the demand for biofuels (although there is a suspicion that felling forest for oil palm is sometimes driven more by the value of the timber extracted to make way for the plantation), spells disaster for the animals that depend on the forest that will be cleared.

All this might - almost - make sense if biofuels really were going to deliver us from climate change, but they're not. Ethanol production from grain requires a lot of energy and, although it does produce less carbon dioxide ($CO_2$) than burning fossil fuels, the difference is only about 10 per cent. Palm oil plantations make a major contribution to deforestation, which itself contributes about a quarter of global greenhouse gas emissions. A back-of-the-envelope calculation suggests that felling rainforest to grow palm oil produces so much $CO_2$ that it would take almost 100 years of burning palm-oil biofuel to get back into 'carbon credit', compared with burning an equivalent quantity of fossil fuel. If the original rainforest were on carbon-rich peatland (which much of it in Southeast Asia is), the payback time rises to 700 years. A sensible attitude to vehicle fuel efficiency could achieve larger reductions while keeping food prices down and not persecuting the world's poor. Biofuels are a symptom of the political need to be seen to be doing something - anything - about climate change, coupled with a deep reluctance to take any really difficult decisions.

Currently the emphasis is on fuel from grain and oil crops, but in the longer term many are pinning their hopes on fuel from biomass, and in particular from crop residues (the non-edible parts of plants left after harvest - essentially straw). The figures certainly look impressive: the total energy value of annual crop residues in the USA alone is estimated as equivalent to almost one billion barrels of diesel. Unfortunately, crop residues already perform other important functions - preventing soil erosion and maintaining soil fertility. Crop residues help to keep up the level of soil organic matter, and within limits there is a direct link between soil organic matter and agricultural productivity, via improvements in soil structure, moisture retention and microbial processes. Ironically, increasing the amount of carbon locked up in soil is also a potentially valuable weapon in the fight against climate change. Put simply, crop residues are too valuable to be burned.

# Climate change

Just when you thought things couldn't get any worse, along comes climate change. Climate change is often bad enough on its own, but its effects are frequently compounded by mismanagement and, sometimes, just plain bad luck. Take the Murray-Darling Basin in Australia, which supports two million people and over a quarter of Australia's agricultural production. After more than ten years of drought, water storage is about 10 per cent of capacity, the iconic river red gums are dying from drought and salinity, and the mouth of the Murray River is kept open only by dredging. Lakes that were crucial habitats for migratory birds are either hyper-saline or have dried up completely. At the time of writing, only heavy rain can save irrigation allocations from being reduced - for the first time ever - to zero.

The Basin has recovered from droughts before, but it happens that the period during which most infrastructure was built and agricultural development occurred (1950-1990) was unusually wet. Things have now returned to drier conditions similar to those that existed during the first half of the twentieth century, but this time round it's a lot warmer and demand for water is much higher. As a result, the resilience of the Basin and its ecosystems is being severely tested, perhaps to destruction.

Globally, climate change looks like quite a different threat from habitat loss and fragmentation, but in terms of impact on biodiversity it's just another form of habitat loss. If a species can no longer survive because its habitat is now too hot or too dry, that's habitat loss for that species, as surely as if the habitat had been burned or bulldozed. Therefore the effects of climate change on biodiversity can be predicted in much the same way as those of habitat destruction, although we need to make two key assumptions.

First, we need to assume that where a species lives now is a good indicator of its climatic tolerance. Although local distributions are often limited by dispersal failure (see Chapter 2), this seems to be generally true at the broad climatic scale. There's plenty of evidence that recent range expansions or declines correspond pretty closely to what one would expect on the basis of recent climate change. What's more, species transported to the other side of the world seem to occupy their new homes in a way that is consistent with their observed climatic tolerance in their original habitat. So, we can use a species' current range to describe its 'climate envelope' - the climatic conditions under which it can currently persist. We can then take the predictions of a climate model for, say, 2050, work out where those climatic conditions will occur, and then compare the geographical location of the future climate envelope with its present location.

Whether the new potential range is smaller, larger or the same size as the present one, one thing is for sure - it won't be in exactly the same place, which brings us to our second assumption. Or rather, a question: can the species occupy its new range? There are two extreme possibilities. We can assume that a particular species is quite unable to spread (or at least, unable to spread fast enough to keep up with the expected rate of climate change), in which case it will survive only where the old range and new range overlap. If there is no overlap, quite rapid extinction is likely. At the other extreme, we can assume perfect dispersal, so that the species is able to spread into all its climatically suitable habitat (within reason). Since in practice we rarely know much about the dispersal ability of individual species, scientists usually report the results of both pessimistic (no dispersal) and optimistic (perfect dispersal) scenarios, even though we know that the reality will be somewhere between the two. Either way, the analysis then proceeds exactly as described earlier for species losses from biodiversity hotspots: compare the area available before and after a certain amount of climate change and, if there's a reduction, calculate the number of species lost.

If we do this, for plants and a range of animals, by 2050 we expect anywhere from 9 per cent (minimum expected climate change, perfect dispersal) to 52 per cent (maximum expected climate change, no dispersal) of species to be lost.[12] When this result was first reported in the journal *Nature*, many people misunderstood this to mean *actual* extinctions by 2050. In fact these are species 'committed to extinction' - as we saw from likely extinctions in hotspots, actual extinctions will take longer than this, perhaps many decades longer. In the end, of course, it doesn't really matter; the key point is that, depending on how optimistic you feel, climate change is either second only to habitat loss as a cause of extinctions over the next 50 years, or the largest single cause of extinction.

It's hard to overstate the uncertainties in these sorts of prediction. They are based only on a very limited range of organisms. With the exception of butterflies, invertebrates hardly get considered at all. Nor has anyone tried to determine how interactions between climate change and habitat loss will affect extinctions, or how many losses due to climate change will be 'additional' to those caused by habitat loss. There will undoubtedly be profound differences between regions, depending on the degree of protection from habitat loss (see below for one example). There's also much uncertainty about the fate of those projected climates that have no present-day analogue. These 'novel' climates tend to cluster in the tropics, often (no surprises here) in biodiversity hotspots. If you think about it, novel climates in the tropics are an inevitable consequence of the broad pattern of climate change, which is a general poleward shift of climatic zones. An across-the-board climate warming means that at the equator, the new climate must be warmer than any existing climate. Not that this is entirely a tropical problem - by the end of this century most of central England is expected to have a climate that doesn't exist *anywhere* in Europe right now. For those ecologists who expect to be around at the time, there will be a certain morbid fascination in seeing what ends up living in future climates that don't exist anywhere today.

In some parts of the world with high rates of habitat destruction, climate change will merely finish off species that would have become extinct anyway. To see the real power of climate change, you have to look at places completely protected from other types of damage. One such location is the upland tropical forest found in a strip of north-eastern Australia around Cairns. This World Heritage area is completely protected and is home to a unique collection of animals and plants: 65 vertebrates are endemic to these relatively cool, moist tropical forests. Dispersal is not an issue here, because as the climate warms, species will have to move only quite short distances uphill, rather than much larger horizontal distances. The problem is the same as for all mountain species, in that keeping up with climate change is easy until you reach the top of the mountain and there's nowhere else to go. The results of climate simulations are clear: any warming over 1°C will cause large numbers of extinctions, and above 3.5°C the job is essentially complete - all the endemic vertebrates will be extinct. Given the predicted temperature increase of 1.4-5.8°C this century, most or all of the endemic fauna of these rich forests is doomed, unless drastic action to curb global warming is taken immediately.

At the risk of being accused of kicking economists again, I have to say that climate change does seem to bring out the worst in some of them. The faith of economists in markets is matched only by their faith in the ability of human ingenuity to develop substitutes for anything. Some even claim that we can afford to ignore climate change, on the grounds that its most severe effects are on agriculture, which accounts for only around 3 per cent of GNP in the developed world. In other words, technology will eventually provide substitutes even for food. Other economists have criticised as alarmist the UK government's 2006 Stern Review on the economics of climate change, which suggests spending 1 per cent of global GNP on reducing greenhouse gas emissions. Given long-term growth in global per capita GNP of around 2 per cent annually, Stern is asking us only to accept the living standards we enjoyed six months ago, which didn't seem too bad at the time.

## Ocean acidification

The main cause of climate change is rising atmospheric $CO_2$ levels, but around half of all the extra $CO_2$ produced so far by burning fossil fuels has dissolved in the oceans. The result is to make the ocean more acid. Seawater is naturally slightly alkaline (about pH 8.2), and the extra $CO_2$ dissolved so far has reduced this by about 0.1. By 2100 ocean pH could fall by 0.5 pH units, which may not sound a lot, but pH is a logarithmic scale, so this value actually represents a three-fold increase in acidity. This pH is lower than has been experienced for hundreds of millennia and, critically, the rate of change is about a hundred times greater than at any time over this period.

The effects of increasing ocean acidity have been so little researched that we have little idea what the impacts will be, but they could be disastrous. The chemistry is complex, but increasing acidity makes it more difficult for animals and plants to make calcium carbonate. All marine organisms with calcareous shells or plates, including molluscs, crustaceans, echinoderms, corals, large calcareous algae, foraminifera and some phytoplankton, will experience problems. Tropical and subtropical corals may be particularly badly affected, with worrying implications for coral reefs. The effect of rising $CO_2$ on climate usually grabs the headlines, but ultimately ocean acidification may be just as bad.

## Focus on birds

It's difficult, perhaps impossible, to put an exact figure on predicted rates of extinction of any group of animals. In fact it's almost impossible even to say how many species are already extinct. The one possible exception is birds, which are diverse (about 10,000 species, more than mammals or reptiles), but not too diverse. They're also well known, relatively easy to find (unlike mammals, few are nocturnal, for example) and, crucially, an extraordinarily large number of people take a keen interest in them. In the UK, the Royal Society for the Protection of Birds has over a million members, more than all the main political parties combined.

So how are birds doing? The first thing to report is that human-caused extinction of birds is not a new phenomenon. Fossil and other evidence suggests that human settlement of the Hawaiian Islands alone led to the extinction of between 70 and 90 endemic birds, long before Europeans arrived. A conservative estimate is that around 1,000 species of bird fell to Polynesian expansion across the whole of the Pacific islands.

Both numbers and timescale are too uncertain to quantify precisely these early extinctions. To express the rate of extinction in historical and recent times, we need a new unit: extinctions per million species-years (E/MSY). The 'background' rate, before humans came on the scene, is about 1 E/MSY. That is, if you followed the fate of a million bird species for one year, you would expect to observe just one extinction. Put another way, our modern complement of 10,000 species should give us one extinction every century. From 1500 to 1800, our best guess is that the rate was about 85 E/MSY (including the world's most famous extinct bird, the dodo). Even though this is much higher than the background rate, it's certainly an underestimate. One reason we know it is an underestimate is that we continue to find evidence of new species that went extinct during this period before they were even described, while some (especially smaller species, unlikely to have left any remains) simply vanished without even being noticed. Early explorers' logs report distinctive island species for which we have no physical evidence or even drawings.

From 1800 onwards we are on firmer ground. Extinctions of birds on islands have continued, but at a lower rate, simply because the most vulnerable species have already gone. On the other hand, birds discovered for the first time in the nineteenth or twentieth centuries tended to have high rates of extinction, because most common species had already been discovered and the new species tended to be both rare and, consequently, more likely to be endangered. As we come closer to the present, the data are much better, but we also have to include the increasing impact of conservation. There have been 20 extinctions in the wild since 1975 (some survive in captivity), but another 25 species, like patients attached to life-support machines, survive only because of intensive conservation efforts. In other words, the underlying, 'real' loss of birds from the wild is 45 species over the last 30 years, a rate of 150 E/MSY.

What of the future? If we apply the usual habitat-loss calculations to the 2,821 bird species endemic to the 25 global hotspots, we expect some 1,700 of these species to go extinct. Several lines of evidence suggest these extinctions might have a half-life of around 50 years, which means that we will lose perhaps 1,250 during this century. Coincidentally, Birdlife International, the global coalition of bird conservation organisations, lists almost exactly the same number of species as threatened with extinction. Both estimates suggest an extinction rate for the twenty-first century of around 1,000 E/MSY.

Bad as this looks, it is in many ways a best-case scenario. It fails to allow for further habitat loss in the hotspots, or extinctions outside the hotspots. If we assume that only areas currently protected will survive in the hotspots, this alone would increase the number of expected bird extinctions to 2,200. It also doesn't consider climate change, which will almost certainly add to the total. For example, the greatest number of threatened bird species in the Americas are to be found in Brazil's Atlantic Forests, where most surviving species occur in upland forests at greatest risk from climate change. Recall the plight of the Queensland forests I mentioned earlier. Finally, birds are prey to a range of other unfortunate 'accidents'. Three-quarters of the world's albatross species are threatened by long-line fisheries, while the introduced brown tree snake has eliminated all the endemic birds from the Pacific Island of Guam since its introduction in the early 1950s.

This brief survey of the state of the world's birds shows just how tough it is to estimate precisely the likely effects of threats to species. Sometimes this uncertainty is used as an excuse for inaction, but in truth ecologists and conservationists are simply trying not to stray too far from the facts. We should also bear in mind that extinction is merely the end of a downward spiral and that dozens of bird species survive only in captivity, that many others are in the 'terminal ward', and that others are 'missing in action'. And, finally, that a smaller proportion of the world's birds are threatened than mammals, reptiles, fish, amphibians or flowering plants, and that few of these groups are likely to receive the attention from conservationists that has so far kept down the 'headline rate' of bird extinctions.

Yangtze river dolphin, *Lipotes vexillifer*

*Chapter 6*

# Are species necessary?

*Such collectors should to a certain extent be regarded as in the same class with those philatelists who achieve a great emotional stimulation from an unusual number of perforations or a misprinted stamp. The rare animal may be of individual interest, but he is unlikely to be of much consequence in any ecological picture. The common, known, multitudinous animals, the red pelagic lobsters which litter the sea, the hermit crabs in their billions, scavengers of the tide pools, would by their removal affect the entire region in widening circles. The disappearance of plankton, although the components are microscopic, would probably in a short time eliminate every living thing in the sea and change the whole of man's life, if it did not through a seismic disturbance of balance eliminate all life on the globe. For these little animals, in their incalculable numbers, are probably the base food supply of the world. But the extinction of one of the rare animals, so avidly sought and caught and named, would probably go unnoticed in the cellular world.*

**John Steinbeck,** *The Log from the Sea of Cortez* (1951)

*This is the most delicate and most important point in scientific studies, to know how to distinguish well what there is of the real in a subject and that which we add to it arbitrarily as we consider it: to recognise clearly which properties belong to the subject and which properties we only imagine it to have.*

**George-Louis Leclerc Buffon,** *Premier discours: De la manière d'étudier et de traiter l'Histoire Naturelle* (1749)

## How many species do we need?

We stand on the crest of a breaking wave of extinctions, looking into an abyss that will wipe out billions of species-years of evolution.

Life on Earth recovered from previous mass extinctions, but it took around 5 million years for diversity to be restored; for all practical purposes, life will never recover. Our grandchildren's grandchildren will look back on the twentieth and twenty-first centuries and marvel at what an interesting place the Earth used to be, and wonder how we managed to screw up so completely, so heedlessly, and so permanently. And yet all of us depend totally, for all the necessities of life, on the life we are squandering.

On the face of it, this looks like an open-and-shut case for conservation, so why isn't everyone signed up? I can't claim a complete answer to this question, but part of the problem is that although humans are good at learning from experience, experience so far teaches us the wrong lesson. And that lesson is that we can get away with business as usual. Throughout human history, we have always been just about as destructive as our technology would allow. The main target of the bone harpoons used by some of the earliest fishermen, 90,000 years ago in what is now the Democratic Republic of the Congo, was a two-metre freshwater catfish. That fish is now extinct, and I think we can all guess why. The Clovis people, the first to tackle the large mammals of the American interior, took a little over 1,000 years to wipe out the mastodon, the giant ground sloth, the giant armadillo, the American camel, and many others. The ancestors of the Maori, finding themselves in the fortunate position of dealing with animals that were completely unused to large terrestrial predators of any kind, took a mere 100 years to exterminate 11 species of large, ostrich-like moas. With hindsight, it's clear that these events were the start of the present human-caused mass extinction, which has been gathering pace ever since.

So if we were to ask: 'Exactly how much of the Earth's biodiversity is actually necessary to keep us in the manner to which we have become accustomed?', the answer has to be: 'Obviously not all of it.' After all, many species are extinct already, and a much larger number are *hors de combat*, and yet we're all still here. More people are sick than should be, and far too many have too little food or fresh water, but that's essentially a failure of mankind to divide up the world's resources equitably, not a sign of immediate problems with the planetary life-support system itself.

# Species and ecosystem services

The conservation case would be immeasurably stronger if it could be shown that, to work properly, ecosystems actually need all of (or at least most of) the species they currently have. In other words, that diverse ecosystems provide food, fuel, fibre and oxygen more efficiently, more reliably and in larger quantities than less-diverse ones. Proving this to be true is an alluring prospect. But there's a problem. You may recall that underpinning the issues discussed in Chapter 2 was the assumption - unspoken, but there nevertheless - that biodiversity is the *outcome* of many other things: climate, fertility, management practices, soil pH, geological history and so on. In other words, that diversity is not an explanation of anything, but is itself something that needs to be explained.

You will also recall that Chapter 2 had a strong botanical bias. That's not because zoologists have worried about diversity any less than botanists, it's just that animal diversity is such a slippery concept. Animals (and remember, most animals are insects) need to be caught, which entails an endless chain of difficulties. Different kinds of trap catch different species, and the same trap operated at different times or in slightly different places often catches wildly different numbers of individuals and species. Moreover, the more individual animals you catch, the more species you catch - strongly suggesting that diversity is more a function of effort expended than an absolute, fixed quantity. There are even statistics to deal with this, which will tell you, after you've caught and identified a million beetles, how many more species you might have found if you'd caught ten million. Not only that, but animals differ enormously in lifestyle, so that predator diversity may not correspond to herbivore diversity, and the pattern of bird diversity may be quite different from that of beetles, while ants may be like neither.

Contrast this with plants. First, there aren't nearly so many. Second, they sit still and wait to be counted. Third, despite their great diversity, they're all fundamentally the same, all doing essentially the same things with the same few basic ingredients: water, light, carbon dioxide and the same few mineral nutrients. (To these practical factors, I could add that plants are also far more important in determining the features of ecosystems that we are most interested in, such as productivity, carbon storage and climate control.) With all these natural advantages, it's hardly surprising that towards the end of the twentieth century, botanists had come up with the most complete and intellectually satisfying theory of biodiversity. A theory firmly rooted in the repeated finding that diversity was the result of, and could be predicted from, other variables. Not that even botanists thought they understood everything - there remains, for example, vigorous disagreement about the origin and maintenance of high tropical diversity - but on the whole they felt they had a pretty good idea of how the world works.

Yet not everyone was happy. Other intellects, cool and unsympathetic, regarded the botanical edifice with envious eyes, and thought they could do better. Many critics were zoologists, some of whom were simply unaware of the progress that had been made by the botanists. Others were immigrants from other disciplines, such as mathematics and statistics. Many were students of population dynamics, a sub-discipline of ecology that had already enjoyed considerable success in explaining pest outbreaks, the cyclic behaviour of some animal populations, and the management of harvested fisheries and wildlife populations. If they could do this, surely they could explain ecosystem function too? Most approached the task with the best of motives, but a few beetle-headed, flap-eared knaves saw an opportunity to put the old-fashioned and unimaginative botanists in their place.

Ironically, wherever they came from, they were compelled to work, initially at least, mainly with plants. There are two very good reasons for this. First, it's easier. Second, plants really are the drivers of ecosystem function. Plants fix the carbon and nitrogen; transpire the water; create and sustain the soil. If you really want to know what makes ecosystems tick, you have at least to start with plants.

They also had another problem. It's clear to everyone that low-diversity plant communities on acid soils (in Europe anyway) didn't make the soils acid in the first place. Equally, low-diversity plant communities on high-phosphorus soils could have played little part in creating the nutrient-rich environment (this is actually an oversimplification in both cases - plants do influence soil fertility - but the fundamental point is valid). If you add fertiliser to a grassland, or increase or reduce numbers of sheep, diversity responds in predictable ways. In other words, in the real world, biodiversity clearly *is* a consequence of other things, rather than the other way round. The conclusion is plain: if you want to demonstrate an *effect* of biodiversity, you need to create artificial communities in which diversity is deliberately manipulated. This is fair enough, and is indeed the classic scientific method of the controlled experiment, in which an attempt is made to keep everything constant except the variable of interest. You can then change that variable alone against a constant background and record what happens. This is not always so simple as it appears. To determine the effect of a particular fertiliser applied as a liquid solution, there must be a control that receives the same amount of water *without* the dissolved fertiliser. Only then can some effect of the water itself - however unlikely - be properly discounted. Similarly, in trials of a new medicine, it's vital that some patients receive a placebo, to demonstrate that just believing you have received a medicine has no effect on the illness.

The key point is that, for the first time, biodiversity itself was to be treated as a classic experimental variable in controlled experiments. Only one hurdle remained to be overcome. It's easy enough to create experimental communities with one, two, four or as many species as you like, but these communities will always differ from each other in other ways too, because - and forgive me if this is too obvious - not all species are the same. In a fertiliser trial, a treatment that receives twice as much fertiliser receives more of exactly the same stuff, but a two-species community doesn't just have twice as many species as a one-species community, it also contains a completely new species. Of course this wouldn't matter if all species were the same, but they aren't. To give a farcical example, you might want to add a new plant to a collection of grasses and see what happens. If the new species is another grass, the effect may be negligible; if it's a tree, the effect will certainly be very large, at least in the longer term.

Those intent on demonstrating the effects of biodiversity quickly grasped that there were two ways round this problem, although neither was a perfect answer. One was to always use rather similar species, so that the absurd tree example couldn't happen. The other was always to make sure that however many species were in the mixture, all of them were also grown on their own. So, if you want to know if a ten-grass community is better than a one-grass version, you grow all ten grasses in monoculture, then if you inadvertently have a 'tree' among your grasses, at least you know about it.

## Biodiversity experiments

By about 1990, the stage was set for the first of a completely new kind of experiment – one that many hoped would show for the first time that biodiversity itself has real, positive, measurable effects on how ecosystems work. Before we go any further, however, and to avoid any possible misunderstanding, note that we are now using the word 'biodiversity' in a very specific sense. In the rather broad sense in which the word is often used (e.g. 'the variety of life in all its forms'), the question 'Does biodiversity have positive effects on how ecosystems work?' is practically redundant. Since biodiversity is simply the living part of ecosystems, the answer to this question must be an emphatic 'yes'. However, there is room for substantial disagreement about exactly which aspects of biodiversity control how ecosystems work. In the traditional view mentioned above, history, climate, soil fertility and other *abiotic* variables combine to determine both biodiversity *and* key ecosystem functions: biomass, productivity, nutrient cycling and so on. This is not to deny that animals and (especially) plants have important effects on ecosystem function, but these effects are largely a consequence of the *kind* of organism present: different environments permit the growth of cacti, grasses, dwarf shrubs, savannah or deciduous or evergreen trees, and it's these different kinds of plant (irrespective of how *many* different grasses, trees or shrubs are present) that determine the kind of ecosystem that develops. The alternative and quite different view of the world (in scientific jargon, a different *paradigm*), while not denying the importance of the kinds of animal or plant present, allocates as much (or even more) importance to the *number* of species present. Specifically: diverse ecosystems (with many species) deliver ecosystem services better than those with fewer species.

The early experiments were spectacularly successful in showing that more-diverse ecosystems do work better, and this early success spawned many imitators who carried out their own experiments, in which they manipulated the diversity of plants, fungi, insects, worms and many other things. Generally, these all showed positive effects too. Typical of these experiments, although much larger than most, was the multinational BIODEPTH experiment, which grew grasslands of varying diversity in eight countries and cost 2 million Euros. At its conclusion in 1999, it was claimed in the press release: "An EU-funded research project has shown that the loss of biodiversity in European grasslands will make them less productive, reducing the amount of energy available to the rest of the food chain and threatening the overall health of the ecosystem." Andy Hector, lead author of the report of the project, said: "In addition to moral and aesthetic reasons to conserve biodiversity, our results now provide strong scientific reasons too. These results provide the type of general ecological principles needed for European conservation and agricultural policies."

Enthusiastic reports of the role of biodiversity in ecosystem function also began to appear in textbooks and in the classroom. Shahid Naeem, one of the people behind the very first biodiversity experiment, illustrates the idea to his students at Columbia University by 'auctioning' a computer, then removing a small part at random before trying to auction it again. Of course, now the computer no longer works properly, illustrating how an ecosystem may be disabled by the loss of species. "I want to show that, like diversity in an ecosystem, you lose function in surprising ways," he says. The larger aim is clear: "If we can bring [the functional significance of biodiversity] into our deliberations, we can speak to decision-makers and policy-makers a lot more clearly than we have been able to in the past."

Another American, Michael Huston, now at Texas State University, was the first to draw attention to the flaws in all these experiments. If the diverse community worked better than the simple one (and it generally did), early enthusiasts for the new experiments assumed this was because of what they called 'complementarity'. The idea is that an ecological community consists of species that fit together like the pieces of a jigsaw. The low-diversity community, like an unfinished jigsaw (or Naeem's computer with a part missing), has holes in it and can't fully utilise all the available space, water, nutrients or whatever.

As more and more species are added, the community gets better and better at doing this. Huston pointed out that because not all species are the same, there was another process at work, now usually called the 'sampling effect'.[1] The principle is that naturally, the species in a mixture will vary in their growth rate, size, competitive ability and their fit to the local environment, and some will be 'better' than others. If species in a mixture are selected at random, it's inevitable that a more-diverse mixture has a higher chance of containing the 'best' species in the available pool. In the ten-grass example, a five-species community has a 50-per-cent chance of containing the best species, while in a one-species community the chance is only 10 per cent. A mixture of all ten species *must* contain the best species, whichever it is. A good analogy is trying to select a pair of socks from a well-mixed drawer full of socks. The more socks you select, the better your chances of getting a matching pair, and if you choose all of them you *must* have a pair, even though there's really no sense in which ten socks are actually better than five - after all, you can only wear two socks at a time.

How can we distinguish complementarity from the sampling effect? Crucially, complementarity and the sampling effect make different predictions about the behaviour of mixtures of species. Complementarity says that all mixtures are better than all monocultures, and the more species you add (up to some limit), the better the result. The sampling effect says that although diverse mixtures are better than monocultures *on average*, there will always be at least one monoculture (the 'best' species) that is as good, or nearly as good, as a high-diversity mixture. We'll look later at which of these predictions turned out to be correct, but meanwhile other scientists were beginning to voice criticisms of the first generation of biodiversity experiments, and some were carrying out their own, more subtle experiments. One thing many people quickly noticed was that increased productivity of mixtures of grassland plants depended heavily on the presence of nitrogen-fixing legumes: in BIODEPTH, the effect of adding red clover to the mixture was four times larger than the effect of doubling diversity. Legumes also turned out to be crucial to increased yield in the flagship American experiment, at Cedar Creek in Minnesota. Adding a nitrogen-fixing legume is essentially adding nitrogen, the benefit of which could hardly be attributed to diversity per se. In any case, the beneficial effect of legumes on pasture productivity was hardly news, and had been known to farmers for millennia.

Others were puzzled that the positive effects of biodiversity only ever showed up in artificial communities synthesised from scratch, which fail to resemble natural ecological communities in one key respect. In real ecosystems, both plants and animals show striking hierarchies of abundance. Put simply, there are nearly always a few very common species and many more rare ones. It seemed reasonable to assume that the common species would have much more influence on ecosystem function than rare ones. Not only that, but countless studies had shown that it was nearly always the rare species that were more likely to go locally extinct. In contrast, the diverse synthesised experimental communities usually had all species equally abundant, at least to start with, and less-diverse versions were created by omitting species at random. To get round these problems, and to try to get a bit closer to reality, some ecologists began to devise 'hybrid' experiments that manipulated natural communities. Studies of American prairies quickly showed two things. First, it mattered a lot whether a community lost rare or common species. In one study, two-thirds of species could be lost without any effect on yield, as long as you started from the bottom of the heap and left the commonest species. In contrast, losing the single most abundant species had enormous effects. Second, you will recall from Chapter 2 that most ecological communities are 'unsaturated', i.e. they don't contain as many species as they could accommodate. It was found that if more species were sown, the diversity of prairies could easily be increased, but yield could not: the new species seemed just to pinch a bit of the space and other resources already used by the current residents.

Furthermore, other ecologists began to show that the properties of real plant communities could be predicted quite well simply from a knowledge of the top two or three most abundant species, irrespective of how many rare species were present. It was these *dominant* species that intercepted most of the light, absorbed most of the carbon, water and nutrients, controlled the quantity and nature of the biomass produced, and thus ultimately had the largest impact on the animals the ecosystem could support.

Finally, after more than ten years of biodiversity experiments, an international team gathered together the evidence from 111 published studies, on organisms ranging from bacteria to beetles, to try to answer the big question: Complementarity or sampling effect? The answer, published late in 2006,[2] was surprisingly clear: whether they looked at aquatic or terrestrial systems; at plants, herbivores or predators; diverse mixtures were uniformly incapable of doing any better than (or in most cases, even as well as) the best single species on its own.

# New theories for old

If you're not interested in theories of how science works, please skip this section, but it's worth briefly pondering how the idea arose in the first place that high local diversity makes ecosystems work better. Supporters of the idea would (and indeed do) portray the debate as a classic clash between opposing world views, or paradigms. In other words, rather like (but much less momentous than) the collision between the ancient view of the universe with the Earth at its centre and the heliocentric Copernican view, or between Newtonian mechanics and the relativistic world of Einstein. It's clear that supporters of the new paradigm (high diversity = better ecosystems) see themselves in the role of Copernicus or Einstein, boldly forging ahead with a new and superior view of how the world works. But there's a problem: as Thomas Kuhn points out in his classic book *The Structure of Scientific Revolutions*, a new paradigm is needed only (and indeed only ever makes any progress at all) when there is something clearly wrong with the old one. Thus the death of an old paradigm is always preceded by increasingly elaborate and implausible attempts to make it fit new observations and experimental results - things that turn out to slot neatly into the new paradigm without any trouble at all. But in this case there is nothing wrong with the old paradigm - that diversity is the result of, and can be predicted from, other variables - indeed, it makes much more sense of the real world than the new one. Instead it's actually the new paradigm that has to be implausibly adjusted to fit the facts.

Faced with the fact that in 44 different controlled experiments on grasslands, in which *everything* else is kept constant and thus the only thing that *can* affect productivity is biodiversity, the most diverse mixtures of species average only 88 per cent of the yield of the best monocultures, what do the supporters of the new paradigm say? Before I tell you, you need to know that the jargon for a mixture that can beat any monoculture is 'transgressive overyielding'. Anyway, if you're sitting comfortably, here's the explanation: "[it has been] shown mathematically that transgressive overyielding requires a greater degree of complementarity than is needed for species to coexist. Thus, even when species exhibit a degree of niche complementarity that is sufficient to stabilise their interactions, this does not ensure that a diverse polyculture will outperform the highest-yielding species. Therefore, one possible explanation for the results of experiments to date is that, although complementarity has been sufficiently strong to generate positive net effects of diversity, it has not been strong enough to generate transgressive overyielding."[3]

Make of that what you will - and no, I'm not sure what it means either - but I hear the sound of trying to have one's cake and eat it. If I had a pound for every impossible result that had been proven mathematically, I would have retired long ago. Note also the weasel phrase 'experiments to date', implying that if we did another 44 studies, the results might be different. Finally, observe that the new paradigm now makes no predictions at all about the relative performance of plant mixtures and monocultures: if mixtures do better, that's evidence for the new theory, but if they don't, well, that's probably evidence for the new theory as well. Heads I win; tails you lose.

To me, that seems to be more or less the end of the academic debate on this topic, although I must emphasise that there are plenty of people out there who think otherwise. However, if you remain to be convinced, here are a few more problems with the idea that ecosystems that are more diverse work better. You've no doubt noticed that I've rather skated round the awkward question of what we mean by 'work'. Some measure of yield or biomass has been the usual yardstick, but there's a whole sub-genre of studies that have looked at invasibility (i.e. resistance to invasion by new species), nutrient uptake, nutrient cycling and other processes, in the hope that more diverse systems would be less invasible, take up nutrients more efficiently, etc. All have eventually reached the same conclusion:

having the *right* species is important, but having lots of species isn't. This effect underpins all the well-known practical examples of the usefulness of diversity. For a productive pasture, grass plus legume is much better than either alone, but that's about as far as it goes. Classic intercropping (e.g. maize and beans) or agroforestry (e.g. ginger beneath coconuts) operate on the same principle and have the same characteristics: combinations of a few carefully matched species, and benefits of diversity that peak at two or three species. A recent large European project has spent a long time and a great deal of money trying to show that growing more than one grass species in pastures might benefit productivity, and has concluded that you might - sometimes - get more grass by growing a mixture of two (or even three) species. Fair enough, but not a finding that is likely to have much impact on farmers' attitudes to genuinely diverse (and unproductive) pastures.

## The questionable value of high productivity

Most studies have focused on productivity, because it's easy to measure, and it's also clear that the rate at which water, $CO_2$ and sunlight are turned into biomass is a rather fundamental property of an ecosystem. But it's far from clear that any particular level of productivity is a 'good thing' in natural ecosystems. If you were being paid to manage a forest to sequester carbon, high productivity might well be desirable, but if you were managing a lake that would be very unlikely to be true (in freshwater, high productivity often leads to algal blooms, anoxia (lack of oxygen) and declines in populations of fish and higher plants). Actually, even the forest example isn't as clear as it looks. A productive forest might soak up $CO_2$ faster than a slow-growing one, but for how long? Globally, there's about four-and-a-half times more carbon locked up in soil than in vegetation, and carbon in soil tends to hang around for longer than in trees, so putting more carbon in the ground might be a more sensible objective.

To look at how we might help to do that, consider the war being fought between sugar maple and hemlock in the forests of eastern North America. It's not a war you would ever notice, but it's deadly serious all the same. Sugar maple is a fast-growing tree of fertile soils, and manages its affairs in the kind of high-turnover, capitalist way of which any sensible American economist would approve. Its leaves are nutrient-rich and, when they fall in the autumn, rapidly break down to recycle the nutrients into fresh tree growth. Accordingly, soils beneath sugar maple are relatively low in carbon. In complete contrast, hemlock is evergreen, grows more slowly, and its leaves (which are low in nutrients and packed with all kinds of nasty things such as tannins and resins) break down only very slowly. Soils beneath hemlock are rich in carbon. To a large extent, the distributions of each tree species depend on soil fertility, with sugar maple on fertile soils and hemlock on less fertile soils. But a lot depends on which arrives first, and each tree is quite good at converting any site into its own ideal habitat: sugar maple makes sure fertile soils stay that way, but hemlock (if it gets in first) can quite quickly lock up the available nutrients in tough, unpalatable leaf litter and turn a fertile site into the infertile one that it prefers. The practical lesson of all this is that slow-growing plants may often provide the ecosystem services we need, even at the price of reducing total ecosystem productivity (in the sense of the total amount of dry matter produced per annum). Soil carbon builds up below slow-growing plants on infertile soils, and much of the world's organic carbon has been accumulated below ground by the activities of low-diversity communities of slow-growing mosses, sedges, dwarf shrubs and evergreen conifers in Arctic and boreal ecosystems.

There's also a deeper lesson here for students of biodiversity and ecosystem function. Those who have attempted to show that high levels of the former are good for the latter sometimes seem to assume that species have some kind of interest in maintaining or generating high productivity. But they don't. Recall the joke about the two hikers who met a bear in the woods. As one turns to run, the other says "There's no point running, you can't outrun a bear," to which the first replies "I don't have to, I only have to outrun you." Success is a relative, not an absolute, commodity. Natural selection favours those species that succeed, by whatever method, and if the price of success is lowering the productivity of the whole ecosystem (as with hemlock), then that's what will happen. Throughout the world, many dry fire-prone habitats are dominated by plants that have evolved to be highly flammable. This is not because being burned is desirable in any absolute sense (for the plant or for the ecosystem), but because it's a good way of getting rid of competitors that are less well adapted to fire than you are. Indeed, one of the painful lessons that northern Europeans had to learn as they colonised the hotter and drier parts of the world was that protecting fire-prone habitats from fire was one of the surest ways of destroying them. Some Californians have yet to learn that chaparral (dry shrubland) burns and that this is not a catastrophe for the ecosystem, even if it is for those who have mistakenly built their log cabins in the woods. High productivity is valuable (to us anyway) in a narrow agricultural sense, but many ecosystems are dominated by species for which raw productivity is far from a top priority.

## What to conserve, and why?

It is far from clear that trying to preserve ecosystem function leads to any very obvious prescriptions for conservation *of species*. Indeed, the ecologists who sought to provide an objective ecological reason for saving biodiversity haven't actually provided much useful practical guidance. The idea that ecosystems just go on working better as you add more and more species may seem like a good argument for conservation, but it suggests only that we should try to conserve everything. The alternative view, that ecosystems depend crucially on relatively few key species, looks better because it holds out the prospect of better targeting of conservation resources. In fact, however, it is equally unhelpful. First, because it's rarely possible to predict in advance what these important species are ("You don't know what you needed till it's gone", with apologies to Joni Mitchell). Second, because as the environment and climate change, the identity of the important species might change (the 'insurance' argument). And third, because we rarely know enough about particular species to target our conservation efforts effectively anyway. In the end, both views amount to a plea to conserve everything, which isn't very helpful.

In 2007 the charity Birdlife International launched a campaign to save the 189 birds in the highest category of threat: 'critically endangered'. These are the birds that we will probably lose in the coming decade if we do nothing. This is a thoroughly worthwhile and praiseworthy campaign, yet you will look in vain for any suggestion that the world will end, or even that anything noticeable will happen, if we do lose all these birds. The birds chosen to be the first to receive help when funds allow are those nearest extinction, and/or those where the help required is most obvious, and where suitable organisations or individuals exist who are well qualified to coordinate and implement conservation action. In other words, conservation priorities are determined by what *can* be saved. Not only are these birds not performing some key role in the ecosystem (as far as we know), but it's hard to see how they could be. Confined to a few kilometres of dune in Brazil, or less than 10km$^2$ of marsh in Baja California, most of these birds are already so rare that they are, in effect, functionally extinct.

Conservationists are, in other words, interested in protecting species from extinction as an end in itself, and any knock-on benefits for ecosystems are a bonus. We shouldn't be surprised: conservation is hard work, underfunded and largely thankless. It's run by enthusiastic people who are passionate about birds, butterflies, orchids or whatever, and for whom saving from extinction the species they love is justification enough. Not surprisingly, the whole debate about biodiversity and ecosystem function seems to have been largely ignored by those at the sharp end of conservation, which in itself tells us quite a lot about conservation priorities. When did you last hear anyone argue that losing the giant panda, the Yangtze dolphin or the mountain gorilla would lead to irreparable damage to ecosystem services? Come to that, when did you last hear anyone complain that things have never been the same since the extinction of formerly common species such as the great auk and the passenger pigeon? When we lose these iconic species, we are all the poorer, but not in any measurable material sense.

## Fixing broken ecosystems

As if that weren't enough, there's another fundamental, practical problem with conserving species as a route to making ecosystems work better, and it's a problem best illustrated by a genuine, ongoing conservation crisis. Britain has 881 species of large moth (and also many small ones). Most of the large moths are rare or local, but 337 are relatively common and widespread; by this I mean that they are common enough for the Rothamsted Insect Survey, which has been trapping them since 1968, to describe the long-term trends in their range and abundance. The results of the survey are chillingly unambiguous: most common British moths have declined in recent decades; some of them steeply. How worried should we be, and what could we do to stem the decline? Since the UK government pledged itself to halt - or at least slow - the decline in Britain's biodiversity by 2010 (a target that everyone now agrees was not met), this is far from an academic question.

The basic problem, as usual, is intensive agriculture. With the loss of flower-rich grasslands, wetlands, hedgerows and managed broadleaved woodland, we have lost the plants that provide moth larvae with food and adult moths with nectar. In other words - a point that cannot be overemphasised - we cannot tackle the moth problem in isolation. Indeed, it isn't a *moth* problem at all. Moths are merely one of the canaries in this particular coal mine, one symptom of entire ecosystems in serious need of repair, and no 'moth-specific' action is likely to be effective; in truth it's hard to know what such action could or would look like. If we repair the damaged ecosystems (less fertiliser and pesticides; more manure and compost, farm ponds and hedgerows) we will help to reduce the declines of everything from plants to birds, via bees, beetles and spiders. Not only that, but we would help to fix many of the other symptoms of damaged ecosystems, from flooding to pollution and climate change, that have little or no direct connection with biodiversity.

You might ask whether the UK 'needs' all its large moths. Well, it's instructive to note that until scientists took a look at the long-term, national data, no one had noticed that our moths were declining at all. Moths (and butterflies and other insects) are so sensitive to climatic fluctuations that even seasoned observers take little notice, as individual species have good and bad years. Dramatic short-term declines in many species are quite normal, almost certainly pass quite unnoticed, and have little or no effect on other parts of the ecosystem. It's certainly quite impossible to say that any of those 337 species are individually of any great importance. Even to attempt, one at a time, to figure out exactly what's gone wrong for all the 200-odd declining species, what the consequences might be, and then what to do about it, would be unthinkable. "We're ten years away from telling policy managers which fraction of species to conserve," remarked one optimistic researcher in 2009, but even if policy managers were waiting for this advice (and trust me, they're not), a much more difficult question is *how* we are supposed to do this. Finally, note that we have no idea what is happening to our *uncommon* large moths (the majority), or what any of them contribute to ecosystem function.

To illustrate a slightly different aspect of the same problem, let's look at another British example: the large blue butterfly (*Maculinea arion*). The large blue has always been an enigmatic butterfly: never common, and with populations that tended to come and go over short periods of time. It also has a very odd life cycle: the larvae feed initially on buds and flowers of thyme, but are then taken into the nests of ants, where they spend most of the next year. It's the only British member of a group of six related species, all with the same odd behaviour. Some species - arguably the better-adapted ones - masquerade as ant larvae and are fed by the ants, but the large blue subsists entirely on a diet of ant larvae. Gradually, during the twentieth century, the details of the large blue's ecology were slowly pieced together, but one crucial fact was discovered too late. Each *Maculinea* species needs a different ant, and the large blue needs *Myrmica sabuleti*, an ant that thrives only on hot, dry slopes with very short vegetation. Not only that, but because the butterfly eats ant larvae, it needs large, thriving ant colonies, as otherwise it runs out of food and starves to death.

Failure to recognise the unique needs of its ant host led inevitably to the failure of attempts to conserve the large blue. In 1930, the committee formed to conserve the large blue established a reserve to protect a large population at The Dizzard, a sea combe on the north Cornwall coast. A local blacksmith was employed to deter collectors, but the established practice of rotational burning of sections of the combe was discontinued in case it was harmful to the butterfly. Predictably, scrub and coarse grasses took over and the large population promptly collapsed, the last butterfly being seen at the site in 1939. Throughout the twentieth century the large blue continued to decline as grazing of its steep hillside habitat became less profitable. Some sites were destroyed by agricultural improvement, others (like The Dizzard) by well-intentioned 'protection'. Eventually, the short turf vital to the butterfly's survival was provided only by rabbit grazing at many sites, and the final straw was myxomatosis. The British race of *Maculinea arion* was declared extinct in 1979.

A full understanding of the large blue's unique ecology came too late to save it, but the reintroduction of butterflies from Sweden has been highly successful, and there are now several thriving populations of large blues in Britain. Is the butterfly now secure? Well, not quite. It's clear that - like many butterflies - it always had a tendency to disappear from established sites and appear at new ones, which is hardly surprising given its extremely exacting requirements. It will finally be safe only when a sufficiently large area of landscape provides a network of sites where accidents of climate or management at any one are compensated by the ability to move to nearby sites. One project, the Atlantic Coast and Valleys Restoration Project, aims to do just that in Cornwall.

It's a fascinating story, with many more twists and turns than I have described, but with two wider messages for conservation. The first is that you'd be surprised how little we know about the exact habitat requirements of most wild animal and plant species. The large blue is a beautiful, iconic and rather mysterious insect, popular with collectors from Victorian times, and it's doubtful whether many other species would have inspired the effort necessary to understand its ecology. Note that 62 species of moth became extinct in Britain during the twentieth century, but I doubt that anyone but a few experts noticed their passing, or has even heard of most of them. Certainly it's difficult to imagine any of them receiving one-tenth of the effort devoted to the large blue. Or that the resources required would have been available in many other parts of the world, with less money and more species to worry about. The second message is that, as usual, protecting the large blue needed a whole ecosystem to be fixed, and that once it was, many other species benefited, including rare plants, several flies, bees, beetles and even a scarce native cockroach. (It's unlikely that anyone would have made the same effort to save any of these other species that was made to save the large blue, or even that some of them would have been missed at all.) This demonstrates the fundamental problem with a conservation approach focused on individual species: namely, you cannot fix ecosystems by preventing extinctions; instead you need to fix the ecosystem, and that will prevent the extinctions.

Just in case you're thinking I've misrepresented how little we know about the biology of rare species by picking on a species as bizarre as the large blue, or if you think there are just too many moths, consider an apparently much simpler and more familiar group: British bumblebees. Unlike moths, bumblebees look manageable - there are only 18 or 19 species in the UK (and only about 300 species in the world). What's more, there's a clear distinction between six species that are common and widespread (indeed practically ubiquitous - you almost certainly have most of them in your garden) and the rest, which are much rarer and mostly also declining (in one or two cases, possibly or certainly extinct). Despite years of research and debate, despite their economic importance as pollinators, despite the existence of a society committed to recording the distributions of bees, wasps and ants in Britain, and despite five species having Biodiversity Action Plans (BAPs) dedicated to their conservation, we really aren't much closer to understanding why the UK is losing its bumblebees. In general the problem is obvious and the same as for the moths - intensive agriculture - but why some species appear immune to its effects while others have collapsed remains almost completely opaque. *Bombus sylvarum*, a species that naturalists hardly bothered recording a hundred years ago ('*Bombus sylvarum* everywhere as usual,' commented one observer in 1928), despite being a priority BAP species, is on the verge of extinction in Britain, and we do not know why.

Bear in mind that this uncertainty persists in Britain, one of the richest countries in the world, with biodiversity both laughably poor and extremely well known by tropical standards. Globally, the detailed knowledge available about British bees, moths and butterflies is something tropical ecologists can only dream about. Many tropical insects are known only as a single specimen, a name and a locality. Many more await discovery, while others are extinct and we never even knew they existed.

# Do we need every species?

You could - wilfully or otherwise - misunderstand the argument I have developed in this chapter, so let me make myself absolutely clear. Am I saying that species are unimportant? No, I'm not - after all, species are the building blocks from which ecosystems are constructed, and we all agree that ecosystems work well only if they contain the right species. But just making sure that certain species still survive - somewhere - is not enough. In the UK, for example, ryegrass monocultures have replaced almost all our species-rich lowland meadows and pastures over the last 60 years. But this didn't happen by some mysterious process of random loss of the original species, a process that can somehow simply be reversed by just putting the missing species back. It happened by wholesale destruction of the original ecosystem, involving ploughing, reseeding, massive fertiliser application, a change from hay-making to silage - in fact, a complete transformation of the British agricultural economy and landscape. The result is a landscape in which the original meadow flowers simply do not and could not survive.

One could argue quite legitimately that the new, fertile landscape created by intensive farming delivers cheap food (for animals and people) in unprecedented quantities, and that this is an important ecosystem service in its own right. So it is, but unfortunately that's *all* it delivers. The challenge is to devise multifunctional landscapes that also deliver better water quality, less soil erosion, more carbon storage and healthier and happier livestock, and are also less dependent on cheap oil and phosphorus. If we can do that, then plants, moths, butterflies, bumblebees and birds will all gain too, but these biodiversity benefits will only be welcome side effects of fixing the fundamental controls on ecosystem functioning.

Of course, if we continue to lose more and more species, bad things will undoubtedly happen as a result of those losses. Take birds, for example, even though they are collectively far less important to ecosystems than plants (or earthworms or bacteria). If we lose birds that disperse plant seeds, many plants will suffer. In the tropics, where birds are important pollinators, the loss of nectar-feeding birds will mean that fewer seeds of many plants will be produced in the first place. If we lose insect-eaters, there will undoubtedly be more pest outbreaks, and if we lose scavenging birds such as vultures, there will be outbreaks of animals such as rats and feral dogs, and of human disease. These are real effects of biodiversity loss, but such *indirect* effects pale into insignificance beside the *direct* effects of the habitat loss and degradation, pollution and climate change that was responsible for the biodiversity loss in the first place. Most bird extinctions will be accompanied (and caused) by the total loss of the ecosystems of which they were a part. Those ecosystems will stop doing what we want them to do, not because they have lost a few species, but because they have simply been swept away, and their biodiversity with them.

We, the ecosystems on which we ultimately depend, and the animals and plants supported by those ecosystems, are all passengers in the same leaky lifeboat. Before we worry too much about how the remaining passengers are going to get along on the voyage, we should make sure we attend to the large and growing hole in the hull. Assuming that if we can only fix the biodiversity crisis, all will be well with the world is (literally, in many cases) to fail to see the wood for the trees. In truth, if we concentrate on fixing the basic fabric of functioning ecosystems, by preserving forests and wetlands, by protecting soils and fixing more carbon, biodiversity will prosper too, without any extra help

If some of this sounds a bit woolly, note that, in practice, smart conservationists have always acknowledged that conserving biodiversity itself is not a sensible route to protecting the ecosystem services on which we all depend, but rather that things are the other way round. Let me give you a practical example. After a tropical forest has been lost to agriculture or logging, what do you do if you want to get it back - quickly? At one extreme you could do nothing, and hope that nature will deliver what we want. Sometimes this works, but if the area to be restored is large, it often doesn't. Alternatively, you could set out to plant all the trees that made up the original forest, which for anything more than a tiny area is impossibly expensive. A compromise, from an Australian idea but most highly developed in Thailand, is the framework species method.[4]

The framework method aims to give natural regeneration a good push by planting just 20 to 30 carefully selected tree species. The chosen trees need to be natives, suitable for the local conditions, easy to propagate, with high survival and rapid growth after transplanting and dense broad crowns to shade out weeds, and attractive to wildlife (in practice, this means they must flower and fruit at a young age). In the seasonally dry tropics, an added requirement is the ability to resprout after fire. Young trees are planted out and cared for intensively (with weeding and fertilising) for at least two years, to make sure they get a good start. The aim is to suppress weeds, re-establish a multi-layered, functioning forest, restore ecosystem processes and improve conditions for the germination and establishment of other (non-planted) tree species. Initially, such a forest is a mere skeleton, with much lower biodiversity than a mature forest, but, crucially, it is capable of attracting seed-dispersing animals that will import seeds of many other species. It is this next generation of trees that will ultimately begin the process of restoring the forest to its original condition. Not only is the framework method proven to work, but it also beautifully illustrates how effective conservation can ignore biodiversity as an objective and concentrate on fixing the basic outline of a functioning ecosystem. Once that's achieved, the biodiversity looks after itself.

## A matter of belief

Earlier in this chapter we saw that many of the pioneers of research into a possible link between biodiversity and ecosystem function were seeking "strong scientific reasons" to conserve biodiversity, which would allow us to "speak to decision-makers and policy-makers a lot more clearly than we have been able to in the past". Failure to find any close relationship between numbers of species and ecosystem function has rather taken the shine off this idea, but I'm not sure it was a realistic objective anyway. I don't claim to know why 'decision-makers and policy-makers' act in the way they do, but I suspect it often has rather little to do with the considered views of their scientific advisors, and rather more with getting re-elected. In fact politicians and businesspeople (and to be fair, the rest of us too) are perfectly capable, like the Queen in *Through the Looking-Glass*, of believing six impossible things before breakfast, if those impossible things are conducive to our continuing wealth, comfort and general peace of mind. Thus, while most of us would agree that burning even those oil reserves we already know about would cause catastrophic climate change, that doesn't stop us searching frantically for more of the black stuff. If scientific evidence counts for anything, why do half of all Americans believe the Biblical calculation that the earth is 6,000 years old, in the face of overwhelming evidence that its true age is nearer 4.5 billion years?

It's interesting to contrast attitudes to biodiversity loss with those to climate change. The British Royal Society noted ruefully in 2007 that, while the climate-change community seems to have been rather successful at communicating its message to the public, "communicating the complexity of the biodiversity issue has not had the same success". Being scientists themselves, they are naturally inclined to attribute the difference to the activities of the scientists of the Intergovernmental Panel on Climate Change (IPCC), and indeed there are tentative moves towards establishing a 'biodiversity IPCC', but I suspect there's more to it than that. From a public relations perspective, the great virtue of climate change is that its damaging effects are immediately apparent - even if not personally, then almost every time you turn on the television or read a newspaper. Of course, not all droughts, floods and fires can be attributed to climate change, but a natural tendency to assign blame somewhere finds a convenient outlet in climate change. In the heat of the moment - as the house goes up in smoke or the floodwaters come in through the front door - most of us would be willing to do almost anything to avert further climate change. Buy a few low-energy light bulbs, postpone the holiday we were planning in Bali, maybe even join Greenpeace. But when the fires have been put out, the water has receded and the mess is cleared up, most of us are just as likely to revert to business as usual, seduced by low taxes, air conditioning and plasma TVs. How much more difficult, then, to stir up any lasting interest in biodiversity loss - a process that is almost invisible and whose effects are apparently negligible?

The message of this chapter can be summarised as follows. Even in highly controlled experiments, the evidence that ecosystems with more species work better than those with fewer species is less than compelling. In the real world, where many other factors with much larger effects on biodiversity are free to vary, the evidence is poorer still. Furthermore, even though it is natural for conservation to focus on saving species, that's not an efficient route to the conservation of ecosystems: there are far too many species, most of them are rare or threatened (not necessarily the same thing), and we often have little idea why, or how to make them less rare or threatened. Only by conserving the basic fabric of the ecosystem and its functions can we hope to provide the conditions under which biodiversity can look after itself. In the next chapter, we will (among other things) take a look at some of the ways that may be achieved.

Large blue butterfly, *Maculinea arion*

## Chapter 7

# Reasons to be cheerful?

*We could never have loved the earth so well if we had had no childhood in it, - if it were not the earth where the same flowers come up again every spring that we used to gather with our tiny fingers as we sat lisping to ourselves on the grass; the same hips and haws on the autumn's hedgerows; the same redbreasts that we used to call 'God's birds,' because they did no harm to the precious crops. What novelty is worth that sweet monotony where everything is known, and loved because it is known?*

George Eliot, *The Mill on the Floss* (1860)

## Maybe things aren't as bad as they seem

Most biodiversity is in the tropics, especially tropical forests, and the direst predictions of global biodiversity loss say that in 50 or 100 years' time only small pockets of tropical forest will survive. However, predicting tropical forest loss is tricky, and many forecasts simply project recent rates of loss into the future. Iconoclasts Joseph Wright and Helene Muller-Landau have questioned whether this is reasonable.[1] Nor are these the usual crisis-deniers and fellow-travellers who basically disagree with any forecast that looks as though it might get in the way of untrammelled economic growth. Wright and Muller-Landau are serious scientists with impressive track records, whose ideas deserve proper consideration.

Their basic argument is that loss of tropical forest has in the past been driven largely by the growth of rural populations who have cleared forest for subsistence agriculture. Because nearly all tropical forest is in developing countries, with high rates of rural population growth, such clearance rates have been high, but these rates of population growth may not continue. The lesson from developed nations is that eventually populations enter a *demographic transition*, from high birth and death rates to much lower values of both, linked to increasing industrialisation and urbanisation and improved education, especially of women. One result is that rural population growth slows dramatically and eventually there may even be major rural depopulation, land abandonment and spread of secondary woodland. Countries that are already well down this road include Italy, Spain and Japan. The effects are clear: over much of the Pyrenees, forest cover has doubled in the past 50 years. In southern France, although development continues along a coastal strip, human populations have collapsed in higher inland areas, with consequent loss of arable land, grassland, olives and vines; all replaced by secondary woodland. One effect has been the return of large predators (sometimes assisted by human introduction) to areas where they have been extinct for generations. Wolves are usually the first, but bears and lynx are not far behind.

Oddly enough, many European observers see many of these changes as a problem: endangering plants and animals of open, non-forested habitats, increasing predation on livestock and possibly (along with climate change) increasing the risk of fire - not to mention ruining the view. Nevertheless, it's clear that demographic transition is an established fact, and that the usual consequence is the spread of trees and large wild animals, both herbivores and predators. If we can expect a similar demographic transition in the tropics, is there a possibility that the tropical biodiversity crisis will largely solve itself? Maybe, but Wright and Muller-Landau's suggestion has provoked a fierce debate about whether the history of the developed world will be repeated in the tropics. Chief among the arguments of the critics are the following. (1) Much of the change in forest cover, as in Europe and North America, will come from secondary forest developed on abandoned farmland. Will this be enough, or do many species depend absolutely on pristine forest? (2) Now that the world is a global marketplace, will rural depopulation really lead to more forest? Or will the demand for meat, soy beans, oil palm, timber, biofuels, oil, gas and minerals simply take over as drivers of forest loss? (3) In some parts of the world, especially Africa, will AIDS, corruption and war postpone the demographic transition until it's too late? (4) Will climate change destroy the tropical forests on its own, rendering all this speculation pointless?

On all these questions, the jury is still out, so right now the answer depends very much on your answer to the old Clint Eastwood question: Do I feel lucky? Even those who hope that Wright and Muller-Landau are right can't help thinking what a shame it would be if we believed them and they turned out to be wrong.

One consequence of a possible expansion of secondary forest in the tropics has been renewed interest in the value of such forests for wildlife. As with many other issues we have explored in this book, you would be surprised how little we know about this. What's more, the little we do know is pretty unreliable, for a variety of reasons. Fortunately, however, there's a massive 'experiment', on the Jari estate in the State of Pará in north-eastern Brazilian Amazonia, which is beginning to provide some of the answers. Around 1,300 km² of the 17,000km² estate was cleared, burned and bulldozed between 1970 and 1980, before being planted with fast-growing tree monocultures, chiefly *Eucalyptus*. Normal commercial practice is to harvest these plantations after five to seven years and then replant, but some harvested plantations were subsequently abandoned, leading to the development of secondary forest. This allowed a large international scientific team to go in during 2004-5 and compare the biodiversity of virgin forest, 14-19-year-old secondary forest and 4-6-year-old *Eucalyptus* plantations. Because one criticism of previous work was the concentration on only a few different kinds of organism, the team made an effort to study 15 different groups, from plants to large mammals, via spiders, lizards and dung beetles.

What did they find? Well, there are many ways of comparing the three habitats, including total numbers of species, how many species typical of primary, undisturbed forest were present, and similarity of species composition, and I won't bore you with the details. In summary, secondary forest, and even the young plantations, were surprisingly good at supporting many primary forest species. Indeed, for some mobile animals such as flies, bees and large mammals, there wasn't much to choose between the three habitats. For other less mobile groups, such as plants, amphibians and lizards (and, surprisingly, birds), secondary forest was less good. On the whole, therefore, an unexpectedly optimistic view emerges of the ability of plantations and young secondary forest to provide conservation services that are complementary to those provided by virgin forest. However, two important health warnings need to be attached to this conclusion. The first is that the secondary habitats were surrounded by intact virgin forest that provided a reservoir of primary forest animals and plants and, however good the secondary forest, there were always some species that were confined to primary forest. Second, we shouldn't be persuaded to put too much faith in plantations. In commercial plantations, invading native vegetation is periodically cleared or sprayed with herbicide, but (for reasons that aren't entirely clear) the plantations studied here had a dense understorey of native herbaceous plants, lianas and small trees. Undoubtedly these were largely responsible for the good performance of the plantations, and we shouldn't expect much biodiversity in really 'clean' commercial plantations.

## Every cloud has a silver lining

Some critics of Wright and Muller-Landau's optimistic prognosis for tropical forests accept that destruction by rural farmers may slow dramatically, and perhaps even go into reverse, but argue that it will be replaced by an even bigger threat, as noted earlier. The global thirst for meat, grain, edible oils, biofuels and oil and gas will mean that tropical forest will increasingly be destroyed by multinational corporations. There's plenty of evidence that this is already happening: cattle ranching in the Brazilian Amazon has tripled since 1990, and there have also been big increases in industrial logging and soya farming. This sounds like bad news - after all, bulldozers can clear forest much faster than laborious hand-clearing, and the roads built as a result will also be exploited by others.

But with the new threat comes an opportunity. The hundreds of millions of rural farmers were scarcely open to influence by developed countries, and the governments that promoted the destruction via loans, tax incentives and road building were also more interested in getting re-elected than in appeasing Western critics. But conserv-ationists can now focus their attention on relatively few large corporations - corporations that have consciences, or at least have shareholders and PR departments who are concerned about their public image. Conservationists have quickly learned the value of targeting corporations that ignore the environment, via public-awareness campaigns and consumer boycotts. Campaigners have already persuaded major retailers in the UK and USA to change their policies to favour tropical timber from sustainable sources. Under pressure from Greenpeace, Unilever, the biggest buyer of palm oil in the world, agreed in 2008 to support an immediate moratorium on deforestation for palm oil in Southeast Asia, and to use its leadership role within the industry to try to persuade all the major players within and outside the Roundtable on Sustainable Palm Oil (including Kraft and Nestlé) to support a moratorium. Using threats of negative publicity, the Rainforest Action Network has also persuaded some of the world's biggest banks to change their lending policies to support more sustainable projects. In 2009, consumer pressure (again coordinated by Greenpeace) persuaded the major shoe companies Adidas, Nike, Timberland, Clarks and Geox to make a commitment not to buy leather from cattle raised on pastures created by Amazon rainforest destruction. Nor is it just a case of punishing those who behave badly: there's also the considerable 'carrot' for industry of growing consumer preferences for eco-friendly products, which may be marketed at premium prices.

There remain very significant challenges. Even companies that try to behave responsibly can find it difficult to trace the origins of many of their raw materials, and there are many opportunities for fraud and corruption. The Forest Stewardship Council (FSC), widely regarded as the gold standard of eco-certification, has not been entirely above criticism. In 2007 the FSC had to withdraw certification from one Southeast Asian company on account of its damaging activities in Sumatra. It's also true that consumer behaviour varies widely across the globe - so far North American and (especially) European consumers have proved much more receptive to environmental concerns than those in Asia. Nevertheless, consumer awareness presents a significant opportunity for influencing commercial activity, and in this conservation interests have a major new weapon in the battle against tropical deforestation, which we can expect to see used with increasing effectiveness in the future.

# Conservation is (sometimes) cheap

As I explained in Chapter 4, conversion of wild ecosystems to human use often fails to make economic sense, once the true costs are counted and once subsidies - hidden or otherwise - are stripped away. Sometimes the costs of habitat conversion are both clear and local, and here the benefits of conservation are particularly obvious. One important ecosystem service that lends itself to valuation is pollination. Many important crops need pollinating insects, and there have been many attempts to value the pollination services supplied by wild habitats. A 2004 study[2] in Costa Rica found that both yield and quality of coffee increased with proximity to the pollinating bees provided by native forest. Coffee plantations more than 1km from forest patches had noticeably lower yields. For the single farm studied, the pollination services of nearby forest patches contributed $62,000 per year in 2000-3, years of rather depressed coffee prices. How does that compare to the other uses to which the forest might be put? As pasture, it would be worth less than its pollination value; if converted to sugar cane it would be worth rather more. In other words, the value of the forest's pollination services alone to a single farm (and there are other farms, and other ecosystem services, such as carbon sequestration and water purification) is about the same as its value if converted to agriculture. Clearly, policies that would allow landowners to capture this value would provide compelling incentives for tropical forest conservation.

Another ecosystem service that can readily be valued is ecotourism. One study,[3] of the Mabira Forest Reserve in southern Uganda, found that visitors to the reserve were strongly influenced by the number of bird species they might see, and also expressed a willingness to pay far more than the present entrance fee for the reserve. The fee is less than $5 (at 2001 prices), but a fee of around $40-50 would be acceptable, and crucially would also be enough (if distributed to surrounding landowners) to alleviate the very strong pressures for agricultural development of the reserve. Because Mabira is close to Kampala, and agricultural rents are high, it represents a very tough test of the ability of ecotourism to compete with agriculture. On the other hand, the reserve is also in one sense a best-case scenario, since the tourism project was funded by the EU and distributes a significant proportion of revenues to local people. This is not always the case; for example, very little of the money from visitors to Chitwan in Nepal or Komodo in Indonesia benefits local residents.

The problem with both of the above examples is that they are far from typical, in that in these cases at least some of the benefits from ecosystem services (pollination or ecotourism) accrue directly to local people, directly compensating them for abstaining from agricultural development. Generally speaking, things are much less equitable. Globally, the costs of not developing strictly protected areas in developing countries (i.e. what you would have to pay people to stop such development, or what economists call 'opportunity costs') are more than $5 billion per year (at 2000 prices). Given that the network of protected areas in developing countries is widely acknowledged to be unsatisfactory, both in funding and extent, the true cost of protecting an adequate network would be at least twice the value quoted. Most of these costs (i.e. the costs of *not* using wild habitats for grazing, firewood or agriculture) are borne by local people, who are generally among the poorest on the planet - a situation that is neither fair nor sustainable.

In contrast, most of the *benefits* of tropical conservation accrue internationally, the largest part being dispersed 'global' services such as carbon sequestration, and massive-but-hard-to-quantify 'non-use' values, such as the value of simply knowing a species or habitat still survives, known as its 'existence value'. Note that these global benefits are large because they contribute to the welfare of very large numbers of people, many of them relatively wealthy. One study in 2010,[4] of the remaining forests in the Eastern Arc Mountains of Tanzania, found that the carbon they contain is alone worth five times their potential value if converted to agriculture. That's before you even start to consider clean water, ecotourism, and medicines and other non-timber products. Clearly, if the local population could actually realise only a part of this value, pressure on these forests (which continue to disappear) would be greatly reduced.

The inescapable conclusion is that those who benefit most from tropical conservation should pay for it, especially since they have the money to do so. The difficulty is in finding a mechanism, and in particular one that distributes compensation both equitably and fairly to all those who incur opportunity costs. One source of funds is increasing payments from membership of NGOs such as the World Wide Fund for Nature (WWF), and donations from wealthy individuals and corporations. These have risen dramatically in recent years, but will only ever contribute a small fraction of the resources needed. Another possibility is to try to bring market forces to bear on the problem. Wealthy consumers may be persuaded to pay a premium for environmentally sound (e.g. FSC timber) or other sustainable or fairly traded products. The growth of 'carbon offsetting' by consumers in wealthy countries funds many tropical conservation projects. Further down the line, protection and restoration of tropical forests may be funded by internationally traded carbon credits. At present this is prevented by major concerns over both compliance and exactly how good some projects are at sequestering carbon, but we are slowly making progress. Already the Kyoto Protocol allocates credits for renewable energy projects and for reforestation, and the aim is to include credits for 'avoided deforestation' in the successor to Kyoto - in other words to pay the owners of biodiversity *not* to do bad things. Another key aim is to make such schemes simple enough to be understood (and used) by small farmers in developing countries: something that's far from the case at the moment.

For the foreseeable future, however, the lion's share of money for conservation in the developing world must come from governments - and ultimately taxpayers - in the developed world. Partly this will require a major re-education of wealthy taxpayers, who must come to realise that the large north-south movements of money required are not in any sense unfair, or even charity in the normal sense, but simply a fair payment for the substantial flow of environmental benefits in the opposite direction. The good news is that the sums involved are not large. Birdlife International reckons that with £19 million (around $30 million) over the next five years, they could save from extinction *all* the world's 189 critically endangered bird species. The total cost of conserving an adequate and representative fraction of wild tropical nature is of the order of a few tens of billions of dollars annually. An effective global network of marine protected areas would cost about the same. I've seen such sums described by conservationists as 'vast', but it's hard to see why. For some reason it's seen as naive to point out that tiny fractions of military budgets could pay for this without anyone really noticing, but I'm going to say it anyway. By August 2007, the cost of the Iraq war to UK and USA taxpayers was estimated at $500 billion. If that's too large a sum to comprehend, consider that every year US citizens spend about $16 billion on Christmas decorations, while UK citizens spend $13 billion on illegal drugs. A good start on solving the most urgent conservation problems could be made with only the $5 billion that Americans spend every year on celebrating Halloween. Finally, I need hardly add that *all* these sums are mere small change compared with the frankly astronomical sums that have been expended in attempts to 'fix' the global economic crisis that started in 2007. So the next time anyone asks you if we really can afford to pay for conservation, you know what to tell them.

## Conservation discovers economics

Since we're talking about money, perhaps it's time to confess that conservationists and academics (i.e. people like me) usually make lousy businesspeople. We may be passionate about birds or butterflies, we may thoroughly understand the science (and sometimes even both), but we don't understand or enjoy wheeling and dealing. This is hardly surprising - if we were interested in business, we would have chosen a different career - but it is unfortunate. Even those of us who take little interest in economics are dimly aware that the returns on the large sums of (mostly public) money spent every year on conservation are often less than impressive.

Here's the fundamental problem: although knowledge is patchy and often inadequate, conservationists are as good as they can be at pinpointing what needs to be conserved. Unfortunately, conservation actions, taken at different times and in different places, can have wildly different costs, and the costs normally vary much more than the benefits. Thus the efficiency of conservation (i.e. objectives achieved per dollar spent) depends far more on getting the economics right than on the actual biology. For conservationists, who are very interested in the latter and not at all in the former, this is a difficult lesson to learn. Not that conservationists are entirely to blame: it doesn't help that economists don't understand ecology either. For instance, consider the idea of 'capital'. In the world of economics, manufactured and financial capital depreciate over time, owing to inflation, technological change and other risks. But 'natural' capital, e.g. forests, lakes, soil and air, has the near-miraculous property of renewing itself, nor does it become outdated or undesirable, and of course it's more-or-less proof against the discovery of cheaper substitutes. In reality, as natural ecosystems are degraded and destroyed, natural capital becomes increasingly scarce and thus worth even more. Thus, while it's perfectly possible to value ecosystems in monetary terms, it's only just begun to dawn on economists that ecosystems do not behave like money, and in fact have almost nothing in common with money or the things we normally buy with money.

If all this sounds a bit abstract, let me give you some examples, starting at the global scale. Anyone who has ever bought a house knows that prices are high where everyone else wants to live, and low where no one wants to live. Conservation costs vary in exactly the same way, and it turns out that there's still plenty of the Earth where no one wants to live.[5] Around half the world's land area can still be described as wilderness, defined as having a human population density of less than 5 people per square kilometre (to put that in perspective, England has almost 400 people per square kilometre). Much of this vast area is also relatively pristine; indeed, we can identify 24 'wilderness areas' that are all at least 70-per-cent intact, together occupy 44 per cent of the global land area, and contain only 3 per cent of the world's human population (only 1.4 per cent when urban areas are excluded).

In case you're wondering where all this wilderness is, the largest areas are the combined boreal forests; Antarctica; the Arctic tundra; the Sahara; Amazonia; the deserts of central Asia, Australia, Arabia and North America; the Congo and the Miombo-Mopane savannas of southern Africa, each of which occupies more than 1 million km². If you look at a picture of the world taken at night, these are the bits that are still almost completely dark. Overall, less than 10 per cent of this area has any legal protection, but protecting the rest would be ridiculously cheap, compared with conservation in the developed world. To take an extreme example, protecting the Russian Arctic would involve annual costs of around ten cents per square kilometre, while the equivalent costs for the UK and USA lie in the range

$5,000-50,000 (at 2000 prices). In fact, conservation throughout much of the undeveloped world is so cheap that although any attempt to arrive at a global figure would be to invite disagreement, there's little doubt that *all* the remaining wilderness could be protected (and remember we're talking here about almost half the planet) for less - probably much less - than the cost of the war in Afghanistan. If that isn't a bargain, I don't know what is.

Why, you may be asking, did nobody tell me this before? Well, there's more than one reason, but a large part of it is an unhealthy preoccupation with biodiversity for its own sake. The wilderness areas and the biodiversity 'hotspots' that I talked about in earlier chapters overlap only a little, and indeed the former contain less biodiversity than you would expect from their huge area (which is basically why they are still wilderness). Only about 18 per cent of plants and 10 per cent of terrestrial vertebrates are endemic to the wilderness areas, and most of these are in just five areas: Amazonia, the Congo, New Guinea, the Miombo-Mopane savannas and the deserts of Mexico and south-west USA. In other words, most wilderness just doesn't have what it takes to attract the attention of those whose main interest is the conservation of biodiversity for its own sake. But this lack of attention is a mistake, for two reasons. First, even if biodiversity levels are low in much wilderness, costs are even lower, so investing in conservation there is still good value in terms of species protected per dollar spent. Second, the wilderness areas have other value: they contain the bulk of the world's biomass, the last intact assemblages of large wild animals, and provide ecosystem services of enormous value, not least the world's largest terrestrial stores of organic carbon. Nor should we forget the moral, aesthetic and spiritual value of wilderness.

As a postscript to the above, you will not be surprised to learn that current spending on conservation fails to reflect the bargains to be had in the undeveloped world. Of the estimated $6 billion spent each year on managing protected areas, less than 12 per cent is spent in the less-developed world. Nor is this just a reflection of lower costs there; estimates consistently show that the shortfall in conservation funding below that actually required is much larger in the developing world.

Fair enough, you may be thinking, but all a bit pie-in-the-sky. In the real world, no one is in a position to take conservation decisions at that global scale. In reality, difficult conservation decisions have to be taken every day at local scales, and it's here that we should really be worrying about getting value for money. I couldn't agree more, and it's in this area that a new breed of conservationists has been making real progress in the last decade. They have begun to realise that if they fail to take account of economics, they risk at least getting poor value for money, and in the worst case actually doing more harm than good. When concern over the conservation of the spotted owl led to curtailment of logging on public land in the Pacific Northwest of the USA, logging and timber production simply moved on to private land instead. This is a specific example of a general problem: purchasing land for conservation nearly always displaces some development on to other land. If the displaced development is on land of higher conservation value, conservation may do more harm than good - something that is particularly likely to happen if knowledge of the distribution of biodiversity is poor and thus the initial investment poorly targeted. Buying reserves can also attract developers keen to capitalise on the nearby presence of the reserve; nothing drives up land prices like knowing that no one can build next door. The fundamental point, now being recognised, is that intervention by conservationists always alters the market for land, changing demand and prices, and effective conservation needs to anticipate these effects.

Of course buying land is a particularly expensive and heavy-handed way to conserve biodiversity. In many parts of the world, much biodiversity is on private land, and the conundrum facing government agencies is how to persuade landowners to behave responsibly for the least cost. In many countries the preferred option is fixed-price payments to landowners in return for undertaking (or refraining from) certain actions. The main UK scheme, the Environmental Stewardship Scheme, is of this type. Such schemes, even if those involved meet all the required targets, have a patchy record of actually achieving the intended biodiversity objectives. For example, a 2001 study[6] showed that the 1.6 billion euros spent on similar agri-environment schemes in The Netherlands have had no beneficial effects on wildlife at all.

One way in which such schemes might be improved is shown by an experiment in 2003 in the Australian state of Victoria. The government knew that there was a lot of important wildlife on private land, but they had little idea exactly where, and even less idea of what it might take to persuade landowners to conserve it. Their innovative solution was to invite bids for conservation payments. All landowners who expressed interest in the scheme were then visited by an ecologist who assessed the property and discussed with the owner the sort of conservation actions that the landowner might take. These included retaining large trees, fencing to exclude stock, controlling rabbits or weeds and so on. After this, most of the landowners submitted sealed bids, containing the proposed conservation actions and an offer price for those actions. The government was now in possession of two crucial pieces of information: (a) a detailed knowledge of both the existing biodiversity value of all the land in the bids, plus the actions proposed to improve that biodiversity and their cost, and (b) the opportunity costs of individual landowners, i.e. the payments they were prepared to accept to curtail their freedom to use their land. Opportunity costs are notoriously variable, and the bidding process revealed whether some sympathetic landowners had very low costs, i.e. were actually prepared to share the cost of conservation. The procedure was then simple: pay the landowners, starting with those proposing the biggest improvements to biodiversity at the lowest cost, until the scheme budget was exhausted. Auctioning conservation payments in this way is remarkably efficient - the budget for the trial was $A400,000, and a simple calculation reveals that to achieve the same objectives with a fixed-price scheme would have cost $A2.7 million - nearly seven times more.

Naturally, this was a one-off trial, and it would be unrealistic to expect such massive savings to persist. For example, if such schemes became routine, one would expect landowners to learn from experience and adjust their bidding accordingly. There are also some important practical issues to resolve. In the trial, the biodiversity assessments were not revealed to landowners for reasons of expense, but in the future they might be. One can envisage pros and cons of such knowledge: a landowner who discovered he was the owner of the last population of a rare possum might decide to raise his bid accordingly, while another might be inspired by such knowledge to participate in the scheme in the first place. Nevertheless, it's clear that the fixed-price schemes that currently dominate the market are not the most efficient way to pay for conservation.

## It's getting harder to destroy the world

The discovery that almost half the world's land surface is still relatively intact wilderness raises many interesting questions, including: Why should this be so? and What will become of these huge areas in the future? In some cases, e.g. Amazonia, it's clear that wilderness survives mainly because we haven't yet got around to destroying it (even if large areas are notionally protected). Generally, however, wilderness is just too useless for agriculture, because it's either too infertile, too cold or too dry (and sometimes all three). Irrespective of its present productivity (which may have been increased or - more likely - reduced by modern land use), the land now used by humans for forestry, crops or grazing started out with a level of productivity between two and four times that of present-day wilderness. In other words, we've already used (and abused) the best bits of the Earth. In fact, it's a matter of some regret that most of the world's cities were established on - and have thus obliterated - prime agricultural land.

By 2040 there may be 8 or even 10 billion people, requiring food production to be increased from 3.5 billion tonnes (in 2000) to 5.5 billion, with a corresponding increase in use of water and nitrogen and phosphorus fertiliser. The increases needed in all three are problematical to say the least. Nitrogen makes up four-fifths of the atmosphere, so at least the quantity of nitrogen available is unlimited in principle, but it takes 873 cubic metres of natural gas to make every tonne of nitrogen fertiliser, which means it's not going to get any cheaper (and it's already too expensive for the world's poor). Phosphorus is second only to nitrogen as a limiting element for plant growth, and in the tropics it's often *the* limiting element. There's no shortage of phosphorus in principle, but reserves of inexpensive rock phosphate could be exhausted before the end of this century. More phosphorus can be found, but it won't be cheap. Agriculture faces a looming phosphorus crisis.

However, water may be the biggest problem. Even today, 2.5 billion people lack regular access to sanitation and fresh water. Already 28 countries are officially classed as water-stressed, and by 2040 this will rise to over 50, home to two-thirds of the world's population. Demands for extra irrigation would be bad enough, but pressure on water now comes from all sides. As countries such as China and India develop and industrialise, one of the first symptoms is greater water consumption. Large parts of Delhi already face massive water shortages, resulting in some bizarre problems. For example, one hospital claims that much of its water supply is being pilfered from its supply pipe before it reaches its destination. Enterprising Delhi residents now routinely fit powerful (and illegal) pumps into their plumbing to coax water into their homes. Some mainline pipes have collapsed from the combined vacuum pressure created by hundreds of such pumps.

Nor is the water you actually use for drinking or washing the whole story, or even most of it. As we become wealthier and eat more meat and processed foods, and acquire more consumer goods, vast quantities of water are needed for their production. It takes about 2,400 litres of water to make one hamburger, and 400,000 litres to make the average car. Every small bag of imported salad from the supermarket exports another 50 litres of drought to the Kenyans who grew it, which is why the water supply to the Masai Mara Game Park is increasingly diverted for agriculture before it reaches the park.

Finally, climate change (quite apart from actual changes in rainfall) is destroying the glaciers that store water during the winter and supply it to cities and agriculture during the summer. Some Andean glaciers are expected to disappear within 15 to 25 years, denying water supplies to major cities and putting food supplies at risk in Colombia, Peru, Chile, Venezuela, Ecuador, Argentina and Bolivia. If you want to see the glaciers in the eponymous National Park in Montana, you're strongly advised to visit before 2030.

Admittedly, this was supposed to be a more cheerful chapter. And in one sense it is - much existing wilderness will now stay that way, simply because we lack the means to convert it to anything else. At least, we won't be able either to live there or grow crops. As meat consumption across the world increases, I wouldn't be surprised to see cattle being raised in some increasingly unlikely places, and of course there's little hope for any wilderness that has the misfortune to sit on top of oil, gas or minerals.

# Biodiversity in farmed landscapes: not all bad news

The direst predictions of biodiversity loss, as discussed in Chapter 5, were based upon a simple assumption: once land is converted to human use, its value for its original complement of wildlife evaporates. This simplifying assumption makes the calculations easier, and in some cases is not far from the truth, but as a general rule is far too pessimistic. So how much wildlife, and what sort of wildlife, can we expect to survive in human-dominated landscapes? There have been surprisingly few attempts to answer this question, and indeed some conservationists would say that even to ask the question is to undermine the motivation to create a comprehensive network of protected areas. But protected areas will always be too few, too small and too isolated to do the job properly. Outside cities, like it or not, we are going to have less wilderness and more 'countryside', and the sooner we know the conservation potential of countryside (and how to improve it) the better.

By far the most complete example comes from a 2003 study[7] centred on Las Cruces Field Station in southern Costa Rica, which has a largely agricultural valley to the north-east and a relatively inaccessible, partially forested ridge to the south-west. Much of the original forest cover was cleared in the 1950s and 1960s, but about 25 per cent remains. About 20 per cent of the cleared land is coffee plantations, about 30 per cent is pasture, and the remainder is bananas, yucca, gardens and fallow. Forest is mostly in small patches, with the field station itself in one of the largest fragments, the Las Cruces Forest Reserve (227ha). A team of ecologists looked at both the birds and mammals that survive in this landscape, but the picture from both was very similar, so we'll stick to the mammals.

The good news is that the majority of the native, non-flying mammals in the region seem to be able to use countryside of one sort or another. Particularly good are coffee plantations next to forest patches, the mammals in which were similar to those found in the large forest reserve itself. The combination of forest and an adjacent crop that provides some structural resemblance to forest (e.g. coffee, which is a shrub) seems to be crucial. Isolated coffee plantations and pasture (whether adjacent to forest or not) still had several native mammals, but were much less good for forest species. What you think of this result depends largely on whether you're inclined to think a glass is half full or half empty. As an example of a New World tropical agricultural landscape, Las Cruces is fairly typical. A more intensive (and more urban) agricultural landscape near the capital San José has fewer birds and mammals, while a similar region in Mexico with crops that provide a better structural analogue of forest (citrus and cacao) has more. Thus many forest animals seem happy with a landscape in which remnant forest patches are joined up by woody plantations into some kind of network.

The bad news is that this is not enough on its own - the forest reserve is essential to maintain the total regional diversity, since several specialist forest species - mammals and birds - were found only there. The worse news is that the largest animals originally present in the region were extirpated long ago, by a combination of habitat loss and hunting (giant anteater, mantled howler monkey, spider monkey, white-lipped peccary and tapir) or persecution (jaguar). The last tapir at Las Cruces was killed in 1970 and the last record of a jaguar, which is considered a pest and systematically destroyed with traps and guns, was in 1973. Thus the surviving mammals, although there are many of them, are generally small or medium-sized. Many of the lost large animals performed important functions in the ecosystem - many amphibians depend on forest wallows created by peccaries - and we have already seen how the disappearance of large predators can have profound effects that cascade right through the ecosystem.

None of this should surprise an observer in the temperate developed world, where similar processes have been under way for millennia. The UK long ago demonstrated that large forest animals and large carnivores are incompatible with a largely deforested and densely settled urban and agricultural landscape. Bear, elk, wolf, aurochs, lynx, wild boar and possibly bison have all been extinct in Britain for at least 400 years, and in most cases for far longer. All are woodland animals and were wiped out by the familiar combination of habitat destruction and hunting. Even though a sound nature-conservation and even economic case can be made for the reintroduction of the beaver and wolf, and despite recent interest in the concept of 'rewilding', reintroduction of some or all of these animals to the UK remains unfashionable and has certainly made little practical progress. Conservation in the UK and other densely populated, developed countries has treated the original 'wild' landscape (or at least most of it) as essentially lost, and has concentrated instead on the wildlife that thrives in managed countryside. Thus the debate in the UK is more about protecting traditional, low-intensity farmland from the twin threats of intensification and dereliction than about the possible re-creation of 'wildwood'. Visitors from the developing world always find it hard to comprehend that British conservation volunteers are at least as likely to spend their days felling trees as they are planting them.

Globally, although it is rarely expressed in these terms, one can characterise the very different European and developing world views as a contrast between two radically different approaches to conservation: on the one hand, wildlife-friendly farming (but perhaps more of it); on the other, increasing intensification of the presently farmed area, allowing more land to be spared from farming altogether. It's far from straightforward to decide which approach will yield the greatest benefits, and in fact the evidence is weak that increasing the intensity of farming on existing agricultural land spares undeveloped land for nature. In practice, what often seems to happen is that the 'extra' land is swallowed up by non-staple 'luxury' crops, grain to feed animals, or biofuels, especially as affluence increases. One of the few things we can say for sure is that although much wildlife can persist in the right kind of farmed landscape, there are always species (in the UK, Costa Rica or anywhere else) that rely entirely on large areas of relatively 'wild' habitat. It's hard to imagine any largely agricultural landscape that would permit the survival of the wolf, tiger, giant panda or orang-utan. Nor does this apply only to large animals; in the UK, loss of woodland means large numbers of beetles that rely on dead wood are now threatened with extinction.

Thus 'wild' reserves, preferably big ones from which humans are largely excluded, must always be one element in a successful conservation strategy. It's the balance between reserves and wildlife-friendly farming that is hard to gauge. Some have suggested that in less-developed countries with substantial wilderness and wildlife as-yet-untouched by farming, sparing as much land as possible by intensifying the rest may be the best strategy. However, others have pointed out that agricultural intensification does not take place in a vacuum. Intensive farming damages larger areas than it actually uses, by diverting water from surrounding areas and increasing pollution from fertilisers and pesticides, and inevitably leads to inward migration, population growth and demand for land for roads and housing. Not only that, but the small farmers who worked the land before intensification are often displaced to other marginal lands. Landscapes such as that around Las Cruces (described on pages 137-8) may not be perfect, but we need policies that actively favour such relatively wildlife-friendly countryside rather than intensive, big-business agroindustry. Experience shows that diverse agricultural landscapes are typically associated with *campesinos* (smallholders) and indigenous farmers. Small farmers are more likely to tolerate neglected, unproductive corners, to apply traditional farming practices and grow traditional crop varieties, to emphasise local knowledge and human labour over fertilisers and machinery, and to grow food for local consumption. Because they understand their land and know how to get the best out of it, they also have *higher* productivity per hectare than large-scale agribusiness.

When considering the balance between intensive and wildlife-friendly farming, however, there are a couple of other things we also need to think about. The first is the assumption (often seen as too obvious to need any actual justification) that there is an unbreakable trade-off between farming intensity and farmed area. That is, if we want wildlife-friendly farming *and* enough food for a growing population, we need to farm a larger area. In fact this trade-off, which seems so obvious when viewed from the developed world, is far from obvious.

The principal objection to low-intensity 'organic' agriculture is that yields are too low. Certainly, yields from organic farming in the developed world are lower than those from conventional, high-intensity farming. In the developing world, however, yields from organic farming are consistently higher than those from conventional farming. The reason is simple: in the developed world, high yields are obtained by application of high rates of artificial fertilisers and pesticides, but farmers in the developing world - few of whom can afford fertilisers or pesticides - can dramatically improve yields by adopting organic practices such as crop rotation, cover cropping, agroforestry, organic fertilisers and better water management. Such practices not only improve crop yields (at least in the medium-to-long term), but they also reduce soil erosion, groundwater contamination, release of greenhouse gases and pest damage, and benefit biodiversity. Nitrogen, the main limiting mineral for crop yields, can be supplied by more (and more intelligent) use of nitrogen-fixing legumes.

Thus models of world food supply suggest that global conversion to organic farming could easily feed the present world population, and even a future larger population, without needing to use any more land.[8] Note that this conclusion does not depend on rose-tinted assumptions of what organic agriculture *might* achieve; it is based simply on existing organic crop yields.

The second point to bear in mind is that, despite my observations earlier in this chapter about the economics of conservation, conservation often has objectives beyond the number of species protected per dollar in a given place. If this were the only metric that mattered, it would be hard to justify *any* of the money currently spent on conservation every year in the developed world, and particularly the money spent in densely populated areas with relatively little biodiversity and high land and labour prices. In fact this expenditure *is* worthwhile, because ultimately biodiversity will survive only if people care about it, and they are unlikely to care about something they never experience. Why would someone who has never even seen a sparrowhawk be expected to care about the fate of Indian vultures or Andean condors? All those who make a special effort to visit a colony of large blue butterflies started out by admiring a small tortoiseshell or a peacock.

In the UK, there is a debate about the value of 'wildlife gardening'. What proportion of Britain's native invertebrates can be expected to maintain viable populations in private gardens? Is it a quarter, a third, or even a half? How far should gardeners go in trying to cater for the needs (by growing particular plants, for example) of rare invertebrates they may never see? Should gardeners make any attempt at all to reproduce 'wild' habitats such as meadows in their gardens? What is the point of the huge sums spent every year on wild bird food, when nearly all of it benefits species that were common already?

Such questions are all very interesting, but ultimately beside the point. The real value of wildlife gardening is that today's youngsters, captivated while still young by the wildlife in their own back yards, will grow up to be tomorrow's naturalists and environmentally aware citizens. A recent ICM poll in the UK found that nearly half of those questioned remembered having their first contact with nature in a garden, while the comparable figure for nature reserves was 6 per cent. In the developing world, the situation is subtly different: not so much a matter of cultivating an awareness of wildlife as trying to instil the right kind of awareness. The team that carried out the Costa Rica study described on page 137 noted that in their study area some wild animals are killed for food and the rest tend to be treated as pests. During their study they saw a river otter, several common opossums and a tayra (a kind of weasel) killed for that reason, despite all three being pretty harmless. We can perhaps hope that a more respectful attitude to wildlife will develop, as it has in much of the developed world, with increasing wealth and better education. Certainly I don't see much future, anywhere in the world, for a conservation strategy whose main plank is separating people from nature.

## Conclusions

Where does all this leave us? Or to put it another way, if you are to take away one idea from this book (and that's surely the most I can expect), what would I like that to be? I think it's simply this: biodiversity loss is not itself a problem, or at least it's not *the* problem. And the loss of specific charismatic species, on which conservation efforts tend to be focused, is certainly not *the* problem. Biodiversity loss is actually a symptom of a deeper crisis and, as any doctor will tell you, treating the symptoms rather than the underlying disease rarely leads to a lasting cure. If this is true, how did we come to be so obsessed by biodiversity for its own sake? The answer is that conservation was born out of crisis and has never shaken off that inheritance. Despite pioneers such as John Muir and Aldo Leopold being interested in wilderness as much for its own sake as for the animals and plants it sheltered, the priority of modern conservation has always been preventing extinctions. Inevitably, therefore, it's always been preoccupied with rarity. In fact, until quite recently, the categories of threat employed by national and international conservation organisations were simply measures of rarity. If a species was rare, it must be in danger and thus worthy of conservation. The paradox is that rarity is commonplace - indeed, almost ubiquitous. If you chose an animal species at random and asked me to guess if it was rare or common, my best strategy would be to guess 'rare' every time (I would be right nearly as often if I guessed 'rare insect'). If we knew just how many species there are, and how many of them are rare, we would be stunned by the sheer size of the conservation task facing us.

Not only is rarity common, but it's the relatively very few common species that make the world go round. Common species make up the structural backbone of ecosystems and determine the quantities and rates of materials processed by those ecosystems. Contrary to the ideas of those who attribute great importance to numbers of species (nearly all of which are, by definition, rare), the important characteristics of different ecosystems can readily be deduced from the traits of a handful of common species. Thus it's the few common species that provide both the framework and resources for the many rare ones, and the basic ecosystem services that are crucial for our continued survival. Much conservation (perhaps most) is driven by the exigencies of crisis management to focus on saving individual rarities, and it's easy to see why. Even if we look only at a few well-documented, large organisms (mammals, birds, amphibians, conifers and a few reptiles), representing a tiny fraction of global biodiversity, we still find 794 species confined to single sites. Two-thirds of those sites have no legal protection, and most face imminent threats from human development. Faced with such a crisis, and the knowledge that all the world's critically endangered birds (for example) could be saved for a relatively trivial sum, it's hard to see how anyone with a spark of compassion could behave otherwise. Yet, far-sighted conservationists have always known that only one strategy has any long-term hope of getting every endangered species off the sick list: to conserve the fabric of whole ecosystems, and let the rare species look after themselves.

Such an approach, essentially based on the value of ecosystem services, has been derided by some critics as 'selling out on nature'. Some things (they say) – like the beauty, cultural importance and evolutionary significance of nature – are just priceless, and we should conserve them for that reason alone. Would civil-rights advocates (they say) have been more effective if they had provided an economic justification for racial integration? This seems a curious argument, since there is of course a powerful economic argument for racial integration, but that's beside the point. Such criticism is founded on far too narrow an interpretation of the economic value of ecosystems. What's more, recent research has shown that conservation projects explicitly aimed at ecosystem services seem able to attract more funding, and engage a more diverse set of funders, than those with more traditional biodiversity aims alone.

If ecosystems were properly valued for what they do, there would be no need to appeal to their moral, cultural, aesthetic and spiritual value. Not that we should ignore these values: there may be room for argument about the sums, but few would disagree with the principle that, say, great works of art should be protected and conserved, and I find it hard to see why wild nature should not be cherished for much the same reasons. No one argues that we can afford to lose the odd Matisse because there are still some left. Moreover, the desire to conserve biodiversity has diverse and highly personal origins, and there are many who respond most strongly to moral or aesthetic arguments. One frustrated conservationist recently remarked (with tongue only partly in cheek) that he hoped the giant panda would hurry up and go extinct soon (at any rate in the wild), then the money wasted on trying to save it could be spent on something useful. But not only is there money available for panda conservation that wouldn't be there for anything else, but such an extinction would be a profound failure of our stewardship of the natural world, and would deliver another small blow to the self-respect and humanity of all of us. Biodiversity loss was hardly uppermost in anyone's mind in the seventeenth century, but I don't think there's any harm in letting John Donne have the last word on the subject:

*No man is an island entire of itself; every man is a piece of the continent, a part of the main; if a clod be washed away by the sea, Europe is the less, as well as if a promontory were, as well as any manner of thy friends or of thine own were; any man's death diminishes me, because I am involved in mankind. And therefore never send to know for whom the bell tolls; it tolls for thee.*

**John Donne,** *Meditation XVII* (1624)

Giant panda, *Ailuropoda melanoleuca*

# Glossary

**Biodiversity**
The variety of life at all levels, from genes to species. Originally a contraction of 'biological diversity'.

**Biodiversity hotspot**
A region that both contains an exceptional concentration of endemic species and is undergoing exceptional loss of habitat. According to the original definition by Norman Myers, a biodiversity hotspot must contain at least 1,500 species of vascular plants (or 0.5 per cent of the global total) as endemics, and it has to have lost at least 70 per cent of its primary vegetation.

**Biofuel**
Any fuel derived in some way from biomass. For example, ethanol made from fermenting sugar or biodiesel from vegetable oils.

**Biomass**
Material from living or recently dead living organisms.

**Boreal forest**
Northern hemisphere coniferous forest, mostly north of 50° north. Boreal forest occupies large parts of Canada, Alaska, Sweden, Finland, Norway and Siberia.

**Carbon sequestration**
The long-term storage of carbon in various forms. Originally applied to storage as organic matter (e.g. wood or carbon in soils, especially peat) or minerals such as carbonates, but now also widely used for techniques such as injecting carbon dioxide into depleted oil and gas reservoirs.

## Community

An ecological community is an assemblage of populations of different species that occur together in the same area. The term may apply to a specific group of species, for example a plant community or predator community, or may include all the organisms in an area.

## Conservation

In the ecological sense, the attempt to protect natural resources, including plant and animal species, as well as their habitat, for the future.

## Dispersal

The movement of individual organisms away from an existing population or away from the parent organism; a process often confined largely to a particular stage in the life cycle, e.g. seeds in plants.

## Ecosystem

The organisms that live together in a particular environment, plus the physical environment with which they interact.

## Ecosystem functions

The exchanges and transformations of matter and energy within an ecosystem. These involve, among other things, decomposition and production of biomass.

## Ecosystem services

Specific ecosystem functions that are beneficial to humans, for example oxygen production, soil formation and water detoxification.

## Endemic

A species or race native to a particular region and found only there.

## Extinction

The termination of any evolutionary lineage, at any taxonomic level (e.g. 'the extinction of the dinosaurs'), but most often applied to species. Extinction of a species can be local, in which case one or more populations are lost but survive elsewhere, or total, in which case all populations vanish. When used without qualification, total extinction is usually intended.

**Habitat**
A particular type of environment, such as a sand dune or temperate woodland, usually in the sense of the typical habitat of a particular species.

**Monoculture**
Usually in an agricultural sense, the planting of one species or variety of crop.

**Natural selection**
The process by which favourable heritable traits become more common in a population over successive generations. The key mechanism of evolution proposed by Charles Darwin.

**Niche**
A vague but useful term: the place a species occupies in an ecosystem - where it lives, what it eats, when it grows and so on. In an abstract sense, a potential role within an ecosystem that a species may or may not have evolved to fill.

**Nitrogen fixation**
The natural process, either biological or abiotic (non-living), by which gaseous nitrogen in the atmosphere is converted into forms (e.g. ammonia, nitrates) usable by living organisms. Much terrestrial biological nitrogen fixation takes place by bacteria that live symbiotically within plants, especially members of the Fabaceae (pea and bean) family.

**Nutrient cycling**
The process whereby chemical elements or molecules are recycled through both biotic (living) and abiotic (rocks, soil, air and water) compartments of the Earth. Important cycles include the carbon, nitrogen and water cycles.

**Opportunity cost**
Essentially, the cost of something foregone whenever an individual or society makes a choice. For example, the foreign holiday you can't afford if you buy a new car instead. In the conservation sense an opportunity cost is the income from, for example, agriculture,

minerals or rent that is foregone if an area of natural habitat is not developed.

## Primary forest

Forest that has never been felled. In practice, since the effects of ancient felling may not be obvious, primary forest is forest that *appears* never to have been felled – that is, forest with mature trees; a multilayered canopy; much large, dead woody material and a characteristic ground flora. It is sometimes called old-growth or virgin forest. Note that in England, 'ancient woodland' means woodland dating back to 1600 or earlier, since before this date planting of new woodland was uncommon, so a wood present before 1600 is likely to be natural. Ancient woodland may have been managed by humans and may have no very old trees; the key characteristic is that the site has been continuously wooded.

## Productivity

Primary productivity is the amount of carbon fixed (incorporated into living material) by plants. (*Gross* primary productivity is the total amount fixed in this way. *Net* primary productivity is this quantity less all the carbon respired by plants.) On land (some marine ecosystems are more productive), annual net primary productivity is highest in tropical rainforest and lowest in extreme deserts, where it may be close to zero.

## Saturation (of a community by species)

The point at which no more species can invade a local community. That is, if new species are added, either naturally or experimentally, they either fail to establish or existing residents are displaced, but the total number of species present does not change.

## Secondary forest

Forest occupying a site that has at some time been artificially cleared of trees, but has reverted to forest either by natural succession or as a result of planting.

## Speciation

The formation of new species, normally dependent on substantial but not necessarily complete reproductive isolation.

## Species
Many definitions exist, but in the usual 'biological species concept' of Ernst Mayr, species are groups of actually or potentially interbreeding natural populations, which are reproductively isolated from other such groups.

## Symbiosis
A mutually beneficial interaction between species. An interaction between individuals of two or more species that results in mutual benefit is called mutualism. Symbiotic mutualism, or symbiosis, is an interaction so close that at least one partner does not normally survive in isolation. An example is the algae and fungi that combine to form lichens.

## Taxonomy
The science (and art) of the classification of organisms. Formerly very much dependent on external features and anatomy, but increasingly using DNA.

# References

## Chapter 1

1. Erwin, T. L. (1982). 'Tropical forests: their richness in Coleoptera and other arthropod species'. *Coleopterists Bulletin*, **36**: 74-5.
2. Owen, J. (1991). *The Ecology of a Garden: The first fifteen years*. Cambridge University Press, Cambridge.
3. Mahdi, A., Law, R. and Willis, A. J. (1989). 'Large niche overlaps among coexisting plant species in a limestone grassland community'. *Journal of Ecology*, **77**: 386-400.

## Chapter 2

1. Wright, S., Keeling, J. and Gillman, L. (2006). 'The road from Santa Rosalia: A faster tempo of evolution in tropical climates'. *Proceedings of the National Academy of Sciences of the United States of America*, **103**: 7718-22.
2. Myers, N. et al. (2000). 'Biodiversity hotspots for conservation priorities'. *Nature*, **403**: 853-8.
3. von Humboldt, A. (1858). *Cosmos: A sketch of a physical description of the universe*. Harper and Brothers, New York.
4. Hewitt, J. E. et al. (2005). 'The importance of small-scale habitat structure for maintaining beta diversity'. *Ecology*, **86**: 1619-26.
5. Janssens, F. et al. (1998). 'Relationship between soil chemical factors and grassland diversity'. *Plant and Soil*, **202**: 69-78.
6. Connell, J. H. (1971). 'On the role of natural enemies in preventing competitive exclusion in some marine animals and in rain forest trees'. In P. J. den Boer and G. R. Gradwell (eds) *Dynamics of Populations*, pp.298-310. Centre for Agricultural Publishing and Documentation, Wageningen.

7. Janzen, D. H. (1970). 'Herbivores and the number of tree species in tropical forests'. *American Naturalist*, **104**: 501-28.
8. Matlack, G. R. (1994). 'Plant-species migration in a mixed-history forest landscape in eastern North America'. *Ecology*, **75**: 1491-502.
9. Crawley, M. J. et al. (2005). 'Determinants of species richness in the Park Grass experiment'. *American Naturalist*, **165**: 179-92.

## Chapter 3

1. Naro-Maciel, E. et al. (2010). 'DNA barcodes for globally threatened marine turtles: a registry approach to documenting biodiversity'. *Molecular Ecology Resources*, **10**: 252-63.
2. Pimm, S. L., Dollar, L. and Bass, O. L. (2006). 'The genetic rescue of the Florida panther'. *Animal Conservation*, **9**: 115-22.

## Chapter 4

1. Costanza, R. et al. (1997) 'The value of the world's ecosystem services and natural capital'. *Nature*, **387**: 253-60.
2. Losey, J. E. and Vaughan, M. (2006). 'The economic value of ecological services provided by insects'. *Bioscience*, **56**: 311-23.
3. Hardin, G. (1968). 'The tragedy of the commons'. *Science*, **162**: 1243-8.

## Chapter 5

1. Hanson, T. et al. (2009). 'Warfare in biodiversity hotspots'. *Conservation Biology*, **23**: 578-87.
2. Clover, C. (2004). *The End of the Line: How over-fishing is changing the world and what we eat*. Ebury Press, London.
3. Lawrence, D. et al. (2007). 'Ecological feedbacks following deforestation create the potential for a catastrophic ecosystem shift in tropical dry forest'. *Proceedings of the National Academy of Sciences of the United States of America*, **104**: 20696-701.
4. Pimm, S. et al. (2006). 'Human impacts on the rates of recent, present, and future bird extinctions'. *Proceedings of the National*

*Academy of Sciences of the United States of America*, **103**: 10941-6.

5. Levy, S. (2006). 'A plague of deer'. *Bioscience*, **56**: 718-21.

6. Nilsen, E. B. et al. (2007). 'Wolf reintroduction to Scotland: public attitudes and consequences for red deer management'. *Proceedings of the Royal Society B: Biological Sciences*, **274**: 995-1002.

7. Sergio, F. et al. (2006). 'Ecologically justified charisma: preservation of top predators delivers biodiversity conservation'. *Journal of Applied Ecology*, **43**: 1049-55.

8. Lawton, J. H. et al. (1998). 'Biodiversity inventories, indicator taxa and effects of habitat modification in tropical forest'. *Nature*, **391**: 72-6.

9. Stevens, C. J. et al. (2004). 'Impact of nitrogen deposition on the species richness of grasslands'. *Science*, **303**: 1876-9.

10. Clark, C. M. and Tilman, D. (2008). 'Loss of plant species after chronic low-level nitrogen deposition to prairie grasslands'. *Nature*, **451**: 712-15.

11. Vittor, A. Y. et al. (2006). 'The effect of deforestation on the human-biting rate of *Anopheles darlingi*, the primary vector of *falciparum* malaria in the Peruvian Amazon'. *American Journal of Tropical Medicine and Hygiene*, **74**: 3-11.

12. Thomas, C. D. et al. (2004). 'Extinction risk from climate change'. *Nature*, **427**: 145-8.

## Chapter 6

1. Huston, M. A. (1997). 'Hidden treatments in ecological experiments: re-evaluating the ecosystem function of biodiversity'. *Oecologia*, **110**: 449-60.

2. Cardinale, B. J. et al. (2006). 'Effects of biodiversity on the functioning of trophic groups and ecosystems'. *Nature*, **443**: 989-92.

3. Cardinale, B. J. et al. (2007). 'Impacts of plant diversity on biomass production increase through time because of species complementarity'. *Proceedings of the National Academy of Sciences of the United States of America*, **104**: 18123-8.

4. Forest Restoration and Research Unit (2006). *How to Plant a Forest: The principles and practice of restoring tropical forests*. Chiang Mai University, Thailand.

## Chapter 7

1. Wright, S. J. and Muller-Landau, H. C. (2006). 'The future of tropical forest species'. *Biotropica*, **38**: 287-301.
2. Ricketts, T. H. et al. (2004). 'Economic value of tropical forest to coffee production'. *Proceedings of the National Academy of Sciences of the United States of America*, **101**: 12579-82.
3. Naidoo, R. and Adamowicz, W. L. (2005). Economic benefits of biodiversity exceed costs of conservation at an African rainforest reserve. *Proceedings of the National Academy of Sciences of the United States of America*, **102**: 16712-16.
4. Swetnam, R. D., Marshall, A. R. and Burgess, N. D. (2010). 'Valuing ecosystem services in the Eastern Arc Mountains of Tanzania'. *Bulletin of the British Ecological Society*, **41**: 7-10.
5. Mittermeier, R. A. et al. (2003). 'Wilderness and biodiversity conservation'. *Proceedings of the National Academy of Sciences of the United States of America*, **100**: 10309-13.
6. Kleijn, D. et al. (2001). 'Agri-environment schemes do not effectively protect biodiversity in Dutch agricultural landscapes'. *Nature*, **413**: 723-5.
7. Daily, G. C. et al. (2003). 'Countryside biogeography of neotropical mammals: conservation opportunities in agricultural landscapes of Costa Rica'. *Conservation Biology*, **17**: 1814-26.
8. Badgley, C. et al. (2007). 'Organic agriculture and the global food supply'. *Renewable Agriculture and Food Systems*, **22**: 86-108.

# Index